DR. DENNIS L. TAYLOR

Minerals, Rocks and Inorganic Materials

Monograph Series of Theoretical and Experimental Studies

6

Edited by

W. von Engelhardt, Tübingen · T. Hahn, Aachen
R. Roy, University Park, Pa. · P. J. Wyllie, Chicago, Ill.

F. Lippmann

Sedimentary Carbonate Minerals

With 54 Figures

Springer-Verlag New York · Heidelberg · Berlin 1973

Dr. *Friedrich Lippmann*
Mineralogisch-Petrographisches Institut der Universität
D-7400 Tübingen

ISBN 0-387-06011-1 Springer-Verlag New York · Heidelberg · Berlin
ISBN 3-540-06011-1 Springer-Verlag Berlin · Heidelberg · New York

Contents

A. Introduction: The Rôle of Mineralogy in the Petrology of Sedimentary Carbonates

By tradition, ancient carbonate sediments have been studied chiefly by paleontologists and stratigraphers. From the study of sedimentary structures and preserved fossil remains, a wealth of information concerning the depositional environments of individual carbonate formations has accumulated. If we disregard acid-insoluble residues, only two rock-forming minerals, calcite and dolomite, occur in abundance in ancient carbonate rocks. Therefore, mineralogy has been of rather marginal importance in traditional studies of sedimentary carbonate rocks. An exception is perhaps the problem of the formation of dolomite, which has always been of interest to earth scientists of various specializations. Because of their mineralogical monotony, sedimentary carbonate rocks have rarely appeared attractive to mineralogists, except perhaps when they have served as the country rocks for mineral deposits.

Nevertheless from time to time, outstanding mineralogists became interested in special aspects of carbonate sedimentology and rendered valuable contributions (see e.g. the papers of ROSE, LINCK, SPANGENBERG, RINNE and BØGGILD contained in the list of references).

The long known occurrence of aragonite in mollusc shells and the discovery of the mineral in recent marine oöids by SORBY were the first indications that younger marine sediments are different mineralogically from their ancient counterparts, and that the diagenesis of calcareous sediments must be accompanied by distinct mineralogical changes.

More recent studies, based on modern mineralogical methods, notably X-ray diffraction, have revealed that unconsolidated calcareous sediments and the biogenic carbonates they contain are characterized by a number of carbonate minerals which do not normally occur in ancient carbonate rocks. Diagenesis of calcareous sediments thus leads from mineralogical variety to monotony, and from this viewpoint the importance of mineralogical factors to carbonate petrology is obvious.

Thermodynamic study suggests itself as the most hopeful approach to an understanding of the conditions of formation and transformation of carbonate sediments. Since minerals correspond very closely to thermodynamic phases, the thermodynamic approach is based on mineralogy as the unifying principle. For example, the stability of the carbonates secreted by molluscs is largely dependent upon the mineral(s) present,

and their identification obviates individual thermochemical studies on
every genus.

The stability relations among sedimentary carbonate minerals are
now more or less well known. The common rock-forming minerals calcite and dolomite are indeed stable phases in the pertinent systems. Most
other carbonate minerals of similar composition which are known to
occur in the younger sediments are metastable with respect to calcite,
dolomite, and magnesite. This implies that the sedimentation of carbonates is determined only in part by stability relations. Kinetic factors,
which allow the formation of metastable minerals, appear to be more
important. Although the diagenetic transformations leading to stable
minerals take place by virtue of thermodynamic requirements, the reactions themselves are triggered by kinetic factors as well. Some of the
reactions leading from metastable to stable carbonate assemblages are
susceptible to simulation in the laboratory; others (e.g. dolomitization)
appear to be so slow that they can be studied only in analogous systems
characterized by reasonable reaction rates. In all attempts to explain the
possible mechanisms of such reactions, we must consider the crystal
structures of the final products as well as of the starting materials. This is
another viewpoint from which mineralogy is important to carbonate
petrology, if we regard the crystal chemistry of minerals as a part of
mineralogy.

A certain parallelism with clay mineralogy suggests itself. In that field
we know much less concerning stability relations than we do in carbonate systems. Evidently for that reason, great emphasis is put on the
crystal structures of clay minerals. In the absence of reliable stability
data, the structures offer a hopeful approach to understanding the conditions of formation and transformation of the minerals themselves. It
must be admitted that this is one of the reasons why the present writer
became interested in carbonate minerals. He wanted to test some of the
lines of reasoning customary in clay mineralogy on the inherently simpler carbonate systems. It is perhaps possible to utilize some of the
experience obtained in carbonate petrology, with due modifications, in
clay mineralogy.

Carbonates are not the most abundant type of sediments; sandy
materials and, in particular, argillaceous sediments are more abundant.
Different estimates are obtained by different approaches for the relative
abundance of sedimentary carbonates (see WEDEPOHL). The total
amount of carbonate constitutes an important loose end in the calculation of geochemical balance sheets (see e.g. MACKENZIE and GARRELS).
Analytical geochemists have supplied us with a tremendous amount of
data on element and isotope concentrations in various carbonate materials. Mineralogy has proved invaluable in rationalizing these data. Hence,

the mineralogical transformations which accompany the evolution of sedimentary carbonates are, at the same time, important geochemical processes.

As far as the petrologically important carbonate minerals are concerned, the present monograph may be regarded as a supplement to existing texts and reference books of mineralogy. In these, the carbonate minerals are treated chiefly with emphasis on isolated and well-formed specimens as originate from mineral deposits, whereas the rock-forming occurrences in the sedimentary cycle are usually mentioned in a cursory fashion. To mineralogists, this is not a serious drawback, since the essential properties of minerals other than crystal shape (e.g. optical data, density) are largely independent of the type of occurrence. However, the non-specialist may receive a biased impression of the quantitative importance of well-crystallized specimen materials.

In conventional texts, crystal structures are usually mentioned and also illustrated as an additional means for the definition of a mineral. However, in view of rapid developments within the past decade, most existing texts are incomplete as far as carbonate structures are concerned. The same is true in relation to the conditions of formation of carbonate minerals. In this field a tremendous amount of information has accumulated and many ingrained views are being seriously challenged.

The first part of this monograph deals with the crystal chemistry of the petrologically important carbonate minerals. Here the crystal structures are discussed and illustrated in some detail. In view of the fact that sedimentary carbonates are frequently characterized by extremely fine grain size, crystal structure, as revealed by X-ray diffraction, is the most convenient, the most reliable, and often the only criterion of identification. Other determinative techniques, both conventional (e.g. optical and staining methods) and modern (e.g. infrared spectroscopy), yield much less detailed information, even where they are applicable.

A detailed grasp of the crystal structures is desirable for understanding a number of petrological mechanisms.

Examples of crystal-chemical rationalization of some such mechanisms may be found in the subsequent sections. Moreover, it is hoped that the crystal-chemical information compiled in this monograph may help students of carbonate petrology in further investigations.

In explaining petrological mechanisms on the basis of crystal structure, the simple theory of KOSSEL and STRANSKI for the growth of ideal ionic crystals is utilized (see CORRENS, pp. 166—170). Simple modifications may be applied to this theory in order to allow a description of crystal growth from aqueous solutions. The explanations here given on the basis of ideal crystals are so qualitative in character that there is little to be gained in attempting to extend such rationalization to imperfect real crystals, where, for example, growth by screw dislocation might be considered (see e.g. CURL, 1962). Present knowledge of crystal imperfections in carbonates does not justify an approach beyond that of the ideal crystal.

In the discussion of the systems $CaCO_3$ and $CaCO_3$–$MgCO_3$ (parts C and D) the emphasis is on mineralogy. Parts C and D may be regarded as an essay on carbonate petrology as seen by a mineralogist. Except for chapter C.II.3. (p. 116ff.) where the different growth forms of oöids are explained on the basis of crystal structure, very little attention is given to microscopic petrography and to larger-scale structures. These aspects of carbonate sedimentology are adequately covered in a number of excellent treatises (BATHURST, 1971; CHILINGAR, BISSELL and FAIR-BRIDGE (editors), 1967; FÜCHTBAUER and MÜLLER, 1970; HOROWITZ and POTTER, 1971).

Reference must be made to these books also for a more complete bibliography on carbonate sedimentology than could be given in the present monograph. Here sedimentological publications have been quoted chiefly for the mineralogical information they contain. In many instances, it was deemed sufficient to quote one or a few characteristic examples in order to illustrate the process or the hypothesis under discussion. The selection was determined chiefly by easy accessibility.

In view of the abundance of sedimentary iron carbonates, a section on systems containing $FeCO_3$ seems to be missing. However, information concerning sedimentary iron-carbonate minerals is rather scanty (see p. 50f.), and not too much work appears to have been done on aqueous systems at lower temperatures (see CORRENS, pp. 272—276, and GARRELS and CHRIST). Iron carbonates do not normally occur together with the more common carbonate rocks, limestone and dolomite, but are found mostly in close association with argillaceous formations. The genesis of sedimentary iron carbonates should be discussed rather in connection with the diagenesis of clays.

To C. W. CORRENS I owe the first indications that mineralogical aspects might be important to carbonate petrology. This impression was enhanced by personal contact with W. F. BRADLEY and D. L. GRAF in 1954—1955. H. FÜCHTBAUER and R. G. C. BATHURST, who taught me much about oöids, B. KÜBLER and notably W. v. ENGELHARDT encouraged me to write on sedimentary carbonate minerals. P. A. WITHERSPOON, C. V. JEANS and K. E. CHAVE read parts, W. v. ENGELHARDT, H. FÜCHTBAUER and W. D. JOHNS read all of the manuscript, which was typed by B. SCHLENKER and EVA FODOR. ERICH FREIBERG expertly prepared most of the illustrations. The advice and help of all is gratefully acknowledged.

B. Crystal Chemistry of Sedimentary Carbonate Minerals

I. Calcite-Type Minerals (Rhombohedral Carbonates)

Calcite, $CaCO_3$, is the most abundant carbonate mineral. This follows from the predominance of limestones over other carbonate rocks. In view of the existence of the other modification of $CaCO_3$, aragonite, such a conclusion is valid only when we know that most limestones consist essentially of calcite. That this is indeed the case had been recognized rather early. One might be tempted to assume that such knowledge was derived mainly from the external properties of the coarser grained limestones. However, in the older mineralogical treatises, e.g. in DANA (1892), very fine-grained rocks, such as lithographic limestone and chalk are mentioned also under calcite. Study of the older publications on the subject (e.g. ROSE, 1856; SORBY, 1879) reveals that these early identifications were based mainly on the correspondence of the densities with that of calcite (2.71), which is distinctly different from the value for aragonite (2.93).

RINNE (1924) was the first to prepare an X-ray powder diagram of Solnhofen limestone as a typical representative of a dense limestone of submicroscopic grain size. The diagram was found to be identical with that of calcite. This was one of the earliest mineral determinations of a fine-grained rock, based on X-ray diffraction and thus on the identity of crystal structure. Since then, the fact that ancient limestones consist essentially of calcite has been confirmed by X-ray diffraction in a large number of instances. Such tests have now become routine, and in many cases, X-ray identifications of calcite in limestones are no longer explicitly mentioned in publications on carbonate petrography, because the result is taken for granted.

The structure of calcite is of interest also because a number of important carbonate mineral constituents of sedimentary rocks, including magnesium- and iron-bearing carbonates, have structures which are identical with or closely related to the calcite pattern. Therefore, the structure of calcite serves as a logical starting point in describing the structures of such minerals. It will be seen that the information thus obtained is essential for intelligent identification and differentiation of rock-forming carbonates. This is especially so in view of the variety of

intermediate members which exist among the calcite-type minerals. Moreover, by focusing on the crystal structures of the minerals, we may
expect to see more clearly the genetic problems concerning the rocks in
which these minerals occur.

1. Calcite and Isotypic Minerals

a) Calcite

The atomic arrangement of calcite was among the first crystal structures to be determined from X-ray diffraction. In his classical paper
W. L. BRAGG (1914) introduced the calcite structure after he had
thoroughly established the structures of a number of minerals of more
simple composition, such as halite (NaCl), sphalerite (ZnS), fluorite
(CaF_2) and pyrite (FeS_2). The relation of the calcite structure to that of
halite was implicit already in this paper, and the kinship of both structures was explicitly mentioned by W. H. BRAGG and W. L. BRAGG (1915)
as an aid to the understanding of the calcite arrangement. Since then it
has become customary, for instructional purposes, to develop calcite
from the inherently simpler halite structure (cf. BRAGG and CLARING-
BULL, 1965).

Conceptually the cube shaped unit cell of NaCl (Fig. 1) is deformed in
such a way that one of its body diagonals is shortened to 76.66% of its
original length while the cube edges are kept straight (Fig. 2). This compression yields a rhombohedron of exactly the same shape as a cleavage
fragment of calcite. The edges meeting at its apex make angles of 101°55′
instead of 90° in a cube. The Na and Cl of the halite structure are then
replaced by Ca and CO_3, respectively. It may be postulated that as many
of the symmetry elements of the original NaCl lattice as are compatible
with the compression along the body diagonal and with the replacement
of Cl by CO_3, are retained in the resulting calcite structure. These are:
the trigonal axis coinciding with the body diagonal along which compression took place, and the binary axes perpendicular to it. In view of
the symmetry and number of components of the CO_3 radical, centers of
symmetry can be retained only for the sites of the Ca cation, whereas
binary axes are possible for the anion sites when these are occupied by
carbon and surrounded by three oxygens which are situated at the corners of an equilateral triangle oriented perpendicular to the trigonal axis.

With the NaCl structure as a starting point, it is thus possible to
develop a plausible model for calcite and for the CO_3 anion. It was
shown by W. L. BRAGG (1914), as well as by later investigators, that the
diffracted X-ray intensities calculated for this model agree with the experimental intensities when a suitable size is chosen for the CO_3 group.
The first value given by W. L. BRAGG (1914) for the C–O distance in

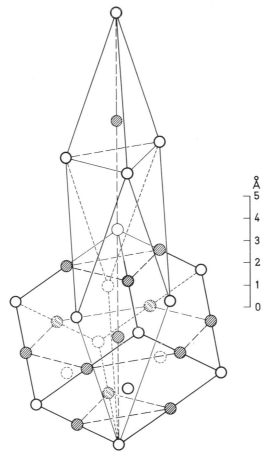

Fig. 1. Cubic unit cell of NaCl (lower part of drawing), containing 4 NaCl. One of the cubic body diagonals is vertical. Octahedral planes with uniform population are thus horizontal and alternate in the direction of the vertical body diagonal. Cl levels are connected by broken lines. The acute rhombohedral cell might be chosen as well for NaCl as a unit cell. However, such a procedure would not be practical because the symmetry displayed by NaCl is not expressed completely ·by this choice. The vertical scale refers to the dimensions of NaCl; open circles: Na; shaded circles: Cl

calcite was about 1.5 Å. Subsequently, W.H. BRAGG (1915) showed that the correct value lies between 1.25 Å and 1.35 Å.

It has been stated above that the calcite cell, obtained by distortion from the face-centered cubic cell of NaCl, has the shape of the megascopic cleavage rhombohedron, i.e. of the conventional reference polyhedron used in morphological crystal studies. In accordance with the

4 NaCl of the original cell, this rhombohedral cell contains 4 CaCO$_3$ as well. By considering that spherical chlorines are replaced by equilateral triangles of CO$_3$, it may be shown that the 4 formula rhombohedron is not a true unit cell. The details of this reasoning are given in the legend to Fig. 2, in which the relation of the cleavage cell to the true rhombohedral unit cell is shown. This latter is characterized by a cell content of 2 CaCO$_3$, and it is described by three reference axes a_{rh}(a_1; a_2; a_3) (cf. Fig. 6) of equal length (6.376 Å). They form the elongate rhombohedron in Fig. 2, and describe angles of 46°4′, contrasted to the obtuse apical angle of about 102° of the cleavage rhombohedron.

In modern crystallographic usage, the symmetry of the calcite unit cell as well as of the whole structure is summarized by the space-group $R\bar{3}c - D_{3d}^6$. This latter was assigned to calcite by SCHIEBOLD (1919) and by WYCKOFF (1920) in their reexaminations of the structure. These authors also pointed out the true rhombohedral unit cell and otherwise confirmed the early structural analyses of the BRAGGs. The equipoints occupied by the chemical components of calcite in the rhombohedral unit cell are given in Table 1a, according to the conventions of International Tables I (1952).

The coordinates of Ca and C are simple fractions of the cell edge. They are said to be situated on special positions. These are determined by symmetry (see column "Point symmetry" in Table 1a). The oxygens, too, occupy special positions in that they are fixed on the binary axes of symmetry. Their exact position along these axes is defined by the parameter x_{rh}, however, in a somewhat circuitous fashion when we use the rhombohedral description. We shall therefore postpone the discussion of the oxygen positions, which have to be deduced from X-ray diffraction intensities, until we have introduced the alternative hexagonal description of the calcite structure. Meantime, the fundamental significance of the rhombohedral description may be appreciated by noting that the symmetry operations which connect the oxygens of the two formula cell

Table 1a. Atomic sites in calcite structure. Space group $R\bar{3}c - D_{3d}^6$ (No. 167; International Tables I)

Occupancy of rhombohedral unit cell

	Wyckoff notation	Point symmetry	Coordinates
2 C	a	32	$\frac{1}{4},\frac{1}{4},\frac{1}{4}$; $\frac{3}{4},\frac{3}{4},\frac{3}{4}$.
2 Ca	b	$\bar{3}$	$0,0,0$; $\frac{1}{2},\frac{1}{2},\frac{1}{2}$.
6 O	e	2	$x,\frac{1}{2}-x,\frac{1}{4}$; $\frac{1}{2}-x,\frac{1}{4},$ x; $\frac{1}{4},$ $x,\frac{1}{2}-x$; $-x,\frac{1}{2}+x,\frac{3}{4}$; $\frac{1}{2}+x,\frac{3}{4},$ $-x$; $\frac{3}{4},-x,\frac{1}{2}+x$.

Coordinates are fractions of a_{rh}.

are expressed by simple cyclic permutations of the coordinates as well as by inversion of their signs (see Table 1a). Moreover, the two-formula rhombohedral cell represents the smallest unit by which the symmetry of calcite and its structure are adequately described, in that it is possible to build up a crystal by filling its space completely with a sufficiently large number of exact replicas of this cell. On account of the small number of atomic sites it contains, it offers many advantages, especially in crystallographic computations, e.g. in the calculation of diffracted X-ray intensities. This becomes obvious when compared to the choice of the cleavage rhombohedron which has to be enlarged to contain 32 $CaCO_3$ in order to comply with the symmetry of the structure (see legend to Fig. 2).

In spite of its arithmetic elegance and simplicity, the description in terms of the two-formula rhombohedron is not simple from the point of view of solid geometry because all angles defining the rhombohedron are oblique. It requires experience and considerable skill in matters crystallographic to build up the structure from unit rhombohedrons and, at the same time, to visualize the details of the structure, such as the mutual coordination of the chemical species. Therefore, an alternative description of the rhombohedral structures is widely used which is based on a hexagonal unit cell. This is defined by a vertical c axis, of the length of the longest body diagonal of the unit rhombohedron, and, perpendicular to this, by two a_{hex} axes of equal length (but different from c) a_1 and a_2, which form an angle of 120°. To these, a third a_{hex} axis, a_3 of equal length, is usually added in a symmetrical position (see Fig. 3), in order to comply with the trigonal symmetry. Different from other crystallographic reference systems, the four-index symbol $h \, k \, i \, l$ is then used for denoting a crystallographic face or plane and the X-ray reflection from the same plane. The "superfluous" index i is not arbitrary but is related to h and k by the equation $i = -(h+k)$ (see e.g. CORRENS, p. 14). Planes generated from a plane $h \, k \, i \, l$ by the threefold axis of symmetry along c are obtained by cyclic permutations: $i \, h \, k \, l$ and $k \, i \, h \, l$. Since the "superfluous" index i is unequivocally determined by h and k, it is often omitted in denoting X-ray reflections, and $h \, k \, . \, l$, or even $h \, k \, l$, is written instead.

For describing atomic sites in a hexagonal system of reference axes, three coordinates $x \, y \, z$ are sufficient as in the other crystallographic sytems. By referring x and y to any two of the axes $a_1 \, a_2 \, a_3$, say a_1 and a_2, (and z to c), we have defined the unit cell on which the description is to be based. For the unit cell chosen, a_3 is then no longer equivalent to a_1 and a_2. If we insist on using a fourth "superfluous" coordinate, this has to be always zero, different from the index i used in describing crystal planes. Hence, the four-coordinate notation for the hexagonal description of an atomic site is written $x \, y \, 0 \, z$ (see International Tables II, 1959, p. 115). The sites $0 \, x \, y \, z$ or $y \, 0 \, x \, z$ obtained by cyclic permutations are

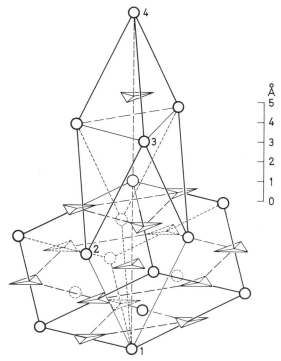

Fig. 2. Cleavage cell or morphological rhombohedron of calcite (lower part of drawing), containing $4CaCO_3$. It has been obtained by deformation of the NaCl cube of Fig. 1. The center of symmetry, coinciding with the Ca at the apex of the cleavage rhombohedron, imposes opposite orientations on the two CO_3 groups which lie above and below, along the vertical body diagonal (broken line connecting 1 and 4). Hence, the 4-formula cleavage cell cannot be a true unit cell of calcite, because its vertical body diagonal represents only half the identity period. The full period, however, is contained in the acute cell with $2CaCO_3$. A true unit cell, shaped like the cleavage rhombohedron, must have a vertical body diagonal of doubled length. Consequently, all of the linear dimensions of the small cleavage cell here presented ought to be doubled. Such a cell is composed of 8 small cells and will thus contain a total of $32CaCO_3$. It is self-evident that such a cell would be rather unwieldy in crystallographic computations. Therefore, the acute cell containing only $2CaCO_3$ is generally preferred in modern crystallography. Of the Ca atoms (open circles) those marked 1, 2, 3, 4 show the interrelation with the hexagonal unit cell in Fig. 3. The vertical scale applies to calcite.
Triangles: CO_3- groups. See also Fig. 6

located in different cells, defined by a_2 and a_3 or a_1 and a_3, respectively, which are generated from the original cell (defined by a_1 and a_2) by the threefold axis along c. However, in contrast to the current use of the four hexagonal indices $h\,k\,i\,l$ for denoting crystal faces and planes, only three

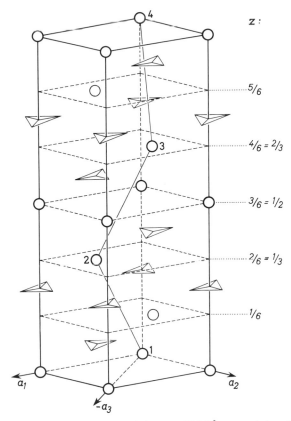

Fig. 3. Hexagonal unit cell of calcite, height: $c = 17.06$ Å, containing $6 CaCO_3$ showing alternating levels of Ca (open circles) and of CO_3 groups (triangles). The z coordinates, in fractions of c, to the right apply to the Ca levels. CO_3 levels are exactly halfway with multiples of $\frac{1}{12}$ as z coordinates. The Ca atoms marked 1, 2, 3, 4 correspond to those in the rhombohedral cell (Fig. 2) marked the same way. 2 haś the coordinates $\frac{2}{3}, \frac{1}{3}, \frac{1}{3}$; 3 has $\frac{1}{3}, \frac{2}{3}, \frac{2}{3}$. These sites are generated from the primitive points 1 or 4 by rhombohedral centering, which makes the hexagonal cell triply primitive

coordinates $x\ y\ z$, parallel to a_1, a_2, and c, are normally used in describing an atomic site with respect to a hexagonal unit cell.

The hexagonal description of the calcite structure is given in Table 1 b. It appears simpler than the rhombohedral choice. The location of Ca and C along the c axis, which coincides with the threefold axis of symmetry, and that of oxygens along binary axes, parallel to the a_{hex}, are recognized immediately for that part of the cell population for which the coordinates are written in full. However, because of the underlying

Table 1b. Atomic sites in calcite structure. Space group $R\bar{3}c - D_{3d}^6$ (No. 167; International Tables I)
Occupancy of hexagonal unit cell

	Wyckoff notation	Point symmetry	Coordinates			
6 C	a	32	$0, 0, \frac{1}{4};$	$0, 0, \frac{3}{4}.$		
6 Ca	b	$\bar{3}$	$0, 0, 0;$	$0, 0, \frac{1}{2}.$		
18 O	e	2	$x, 0, \frac{1}{4};$	$0, \quad x, \frac{1}{4};$	$-x, -x, \frac{1}{4};$	
			$-x, 0, \frac{3}{4};$	$0, -x, \frac{3}{4};$	$x, \quad x, \frac{3}{4}.$	
			$+ (0, 0, 0;$	$\frac{1}{3}, \frac{2}{3}, \frac{2}{3};$	$\frac{2}{3}, \frac{1}{3}, \frac{1}{3})$	

The first two coordinates are fractions of a_{hex}, the third is referred to c. Only that part of the coordinates which corresponds to a primitive hexagonal cell is written out. The coordinates ensuing from rhombohedral centering are obtained by adding corresponding coordinates of the sets in parentheses to corresponding primitive coordinates.

$x_{(rh)}$ and $x_{(hex)}$ are related by $x_{(rh)} = x_{(hex)} + \frac{1}{4}$.

rhombohedral pattern, the hexagonal cell is not primitive. Its full set of atomic sites (or equipoints) is obtained by vector addition (addition of corresponding coordinates) of the coordinates written out in Table 1 b to the coordinates of the points $\frac{1}{3} \frac{2}{3} \frac{2}{3}$ and $\frac{2}{3} \frac{1}{3} \frac{1}{3}$. It is with respect to these points that rhombohedral centering then takes place (for details see BUERGER, 1942 or SANDS, 1969). This operation results in a content of 6 CaCO$_3$ for the hexagonal cell. It is represented in Fig. 3 where also the relation to the rhombohedral cell is indicated. A number of structural details are now seen more directly than is possible in the rhombohedral cell. Inspection of the z-coordinates shows that calcium and carbonate are segregated in alternate levels, perpendicular to the c-axis and spaced at $\frac{1}{12}$ of its length. Consequently, equally populated layers are spaced at $\frac{c}{6}$. However, as a consequence of the rhombohedral centering, these layers are shifted and rotated with respect to each other in such a way that the full identity period is c. On returning once more to our starting point, the NaCl structure, we see that the basal Ca and CO$_3$ layers of calcite correspond to the octahedral Na and Cl planes, which are perpendicular to the vertical body diagonal in Fig. 1.

Identical z coordinates in Table 1 b for adjacent carbons and oxygens are indicative of planar carbonate units. The oxygens are located along binary axes, and their x-parameter directly measures the distance C–O from the nearest carbon. The unit is a_{hex}, so that $C–O = a_{hex} \cdot x_{hex}$. The range for the C–O distance determined by BRAGG (1915) and confirmed shortly afterwards by other investigators (see above) remained unchallenged for more than forty years. Determinations of the C–O distance for

calcite which conform with modern standards of crystal structure analysis have been carried out independently by INKINEN and LAHTI (1964)
(referred to below as IL) and by CHESSIN, HAMILTON and POST (1965)
(CHP below). Both groups of authors used counter techniques for the
determination of diffracted X-ray intensities. IL obtained them from a
powder whereas CHP worked on single crystals. In view of this difference in procedure, it is remarkable that both groups arrived at closely
agreeing results for x_{hex}: 0.2567 and 0.2570, respectively. Moreover, CHP
found no significant change for x_{hex} with temperature in measurements
down to 130° K. The value of 1.279 Å given by IL seems low in comparison to 1.283 Å of CHP. This is explained by the use of obsolete values for
lattice parameters in the paper of IL. When referred to the value of CHP,
$a_{hex} = 4.990$ Å, which agrees with more recent determinations (e.g. GRAF,
1961), the parameter $x_{hex} = 0.2567$ of IL yields 1.281 Å for the C–O distance, and the rather small deviation from 1.283 Å of CHP may be considered to reflect the limit of error for modern parameter determinations.

The investigations of IL and CHP were mainly concerned with the
thermal motion of the atomic components in the calcite structure. Although the two groups of authors used different notations for the thermal parameters, which are thus not directly comparable, there is agreement that the maximum vibration amplitude of oxygen is inclined about
47° with respect to the c-axis. In view of this special scope of their
research, the authors did not utilize their positional parameters x_{hex} to
deduce mean interatomic distances other than C–O. Such an evaluation
has been carried out for this monograph, and the results are contained in
Table 2.

For their discussion, it is useful to consider first the coordination
properties of the calcite structure. In the NaCl type structure each Na
has six immediate Cl neighbors, and vice versa. They are located at the
corners of a regular octahedron. When we use NaCl for a starting point
as done above, it follows for calcite that each Ca is surrounded by six
CO_3 as well. The resulting octahedron with the carbons of CO_3 at its
corners is flattened in the direction of the hexagonal c axis, i.e. in the
direction perpendicular to two of its faces, in accordance with the distortion of the cleavage rhombohedron in comparison to the NaCl cube. The
edge lengths of the flattened octahedron which lie in the basal plane
coincide with the a_{hex} length of 4.99 Å (see Fig. 4) whereas the edges
connecting carbons of adjacent basal carbonate levels are shorter,
4.048 Å. The carbons, however, are not the nearest neighbors of calcium.
The oxygens of CO_3 are much closer and thus form the immediate
neighbors of calcium. This is seen in Fig. 4, which shows the projection of
a basal Ca layer with adjacent CO_3 layers. In the CO_3 layer above the Ca
level the oxygens are situated on the a_{hex} axes and form equilateral trian-

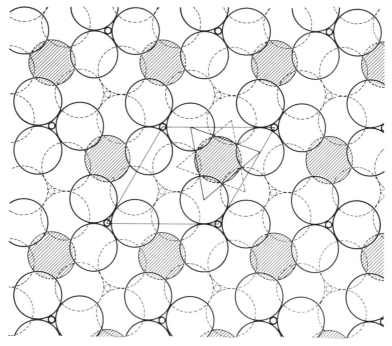

Fig. 4. Individual calcium carbonate layer of calcite structure projected in the direction of the hexagonal c axis, showing lozenge-shaped outline of hexagonal unit cell, edge-length $a_{hex} = 4.99$ Å. The Ca atoms (shaded circles) have the z co-ordinate $\frac{1}{6} = \frac{2}{12}$. The CO_3 layer on top (heavy open circles) has $z = \frac{1}{4} = \frac{3}{12}$ and that below (broken circles) $z = \frac{1}{12}$. The Ca within the outline of the unit cell shows coordination octahedron formed by the centers of surrounding oxygens. Only the equilaterally triangular faces of the octahedron parallel to the base have been drawn. The isosceles triangles inclined to the base remain unconnected. The Ca of the level which would follow on top of the uppermost CO_3 level would be located at the centers (C atoms) of the lowermost and vice versa

gles around the projection of Ca. Oxygen triangles of equal size in the CO_3 layer below are inverted with respect to the ones above by the centers of symmetry in the Ca. Thus two triangles, in levels above and below, describe an almost regular octahedron around each calcium, and it is easy to see that the six oxygens at the corners of the octahedron belong to six different CO_3 groups. In contrast to the flattened CO_3 octahedron with C at the corners, the oxygen octahedron is slightly elongate in the direction of the c axis: its edges parallel to the basal plane (3.26 Å) are shorter than those (3.41 Å) inclined to the base. The Ca–O distance, from the center of the coordination octahedron to its corners, is

2.36 Å. This is close to the sum of the ionic radii of Ca and O: 2.38 Å (GOLDSCHMIDT) or 2.39 Å (PAULING). It may be concluded that the bonding between Ca and the oxygens of CO_3 is mainly ionic in character. The Ca–O length is at any rate considerably greater than the C–O length. In view of this distance relation it appears justified to regard the CO_3 groups or ions as discrete entities, and moreover to relate the calcite arrangement to the ionic structure of NaCl. When regarding the Ca–O bond from the point of view of the CO_3 groups, we find that an individual oxygen is coordinated by two Ca, above and below, which are related by the binary axis through the oxygen and the carbon center.

The O–O distance of 2.22 Å within a CO_3 group is less than two ionic radii of oxygen: 2.64 Å (GOLDSCHMIDT) or 2.80 Å (PAULING). This indicates that the bonding within the CO_3 may be only partially ionic in character. The effective charges must be smaller than the maximum charges of O^{2-} and C^{4+}, which would apply to a purely ionic bond. The situation may be equally well described either by an amount of covalency, active along with the ionic forces, or by the polarizing action of the highly charged C^{4+} ion on the electronic systems of the O^{2-} ions. In 1955, GRAF and LAMAR reviewed the speculations along these lines which had been published up to that time. A reappraisal of the problem on the basis of the new reliable C–O distance appears desirable.

An interesting investigation, which is rather independent of the measured C–O length, is that of LENNARD-JONES and DENT (1927). These authors used a purely electrostatic model and assumed maximum charges for the components of CO_3. A covalent portion for the C–O bond is indicated by a distance between the charge centers of the CO_3 group, of 1.08 Å, which is shorter than the value determined from X-ray diffraction. By expressing the crystal energy ("lattice" energy) as a function of the edge length of the rhombohedral cell and of the rhombohedral angle, the authors were able to show that the values for the lattice parameters which correspond to the maximum structure energy of 701 kcal/mole (771 kcal/mole for the isotype $MgCO_3$) are quite close to those actually observed. Thus the rhombohedral angle occurring in calcite (and magnesite), that is the distortion of the morphological cleavage rhombohedron in comparison to the NaCl cube, affords the most stable structure. The realistic results, obtained from the electrostatic structure model, may be taken as evidence for the generally accepted view that the structure of calcite is indeed held together by the electrostatic forces between the component ions Ca^{2+} and CO_3^{2-}.

This view is further substantiated by the perfect cleavage of calcite along the faces of the morphological rhombohedron. This property may be understood according to J. STARK (1915) by considering that the atomic planes parallel to the cleavage faces are occupied by oppositely charged ions. This situation follows from the kinship with NaCl, and the electrostatic interpretation was originally given for the cubic cleavage of this structure. Adjacent atomic planes parallel to the cleavage rhomb are held together by juxtaposition of oppositely charged ions. By way of some severe deformation, two of these planes are thought to be displaced with respect to each other to such an extent that equally charged ions come into juxtaposition. The resulting electrostatic repulsion then destroys the coherence of the crystal parallel to these planes. The cleavability of other ionic crystals (e.g. fluorite, CaF_2) is equally well explained by the same mechanism (CORRENS, 1969, pp. 108—110). This way, by its cleavage, calcite is related to a number of crystals for which ionic bonding is generally accepted.

It is interesting to note that the glide twinning of calcite (see Fig. 9), which is observed in thin sections of deformed carbonate rocks as frequently as the cleavage, is also amenable to an explanation in terms of ionic forces (BRADLEY, BURST and GRAF, 1953, see under dolomite).

b) Isotypes of Calcite

Long before the discovery of X-ray diffraction, crystallographers had observed that the carbonates of certain bivalent metals other than Ca (see Table 3) are very similar to calcite, in that they exhibit the same morphologic symmetry, of the crystal class $D_{3d} - \bar{3}m$, and very similar rhombohedral cleavage angles as well as optical properties. It was concluded that calcite and these other rhombohedral carbonates are isomorphous. The early X-ray studies of such minerals as rhodochrosite,

Table 2. Interatomic distances (in Å) in rhombohedral carbonates and related cubic oxides XO

		CaCO$_3$ calcite	MnCO$_3$ rhodochrosite	FeCO$_3$ siderite	MgCO$_3$ magnesite
In basal plane	C–O	1.281(3)	CO$_3$ group:	Calculation from C–O value of calcite:	
	O–O	2.219(21)			
	O–O	3.263(1)	3.064(2)	2.981(0)	2.930(28)
(cubic oxide)	(O–O)	(3.401)	(3.144)	(3.046)	(2.979)
Out of basal plane	O–O	3.411	3.153	3.087(6)	3.021(0)
	X–O	2.360(59)	2.198(7)	2.146(5)	2.104(3)
Sum of GOLDSCHMIDT ionic radii	X + O	2.38	2.23	2.15	2.10
(cubic oxide)	(X–O)	(2.405)	(2.222)	(2.153)	(2.106)

Based on the lattice constants of calcite $a = 4.9899$; $c = 17.064$ for 26° C (GRAF, 1961) and the parameter of INKINEN and LAHTI (1964) $x = 0.2567$. The last digits of the distance deduced from the parameter of CHESSIN, HAMILTON and POST (1965) $x = 0.25706$ follow in parentheses where they differ.

$MnCO_3$, and siderite, $FeCO_3$, revealed that they yield diffraction features which are analogous to calcite (BRAGG, 1914; WYCKOFF, 1920). These minerals were thus found to be indeed isostructural with calcite. The carbonates known to date to be isotypes of calcite on the basis of their X-ray diffraction patterns are listed in Table 3 together with their lattice dimensions. WYCKOFF (1920), who incidentally was among the first authors to expound the importance of ionic bonding in calcite, was able to show that the C–O distance in rhodochrosite and siderite agrees closely with that in calcite, "though other distances between atoms vary quite widely". This view is now generally accepted, although modern parameter determinations for simple rhombohedral carbonates other than calcite are lacking. We do have, however, a rather reliable value for the C–O distance in dolomite, of 1.281 Å (see below), which is the same as that for calcite within the limit of error. Therefore, in the absence of more immediate information, we may for the moment assume that the size of the CO_3 group in rhombohedral carbonates is independent of the cation. The interatomic distances for the structures of the rock-forming carbonates rhodochrosite, siderite and magnesite have thus been calculated from the C–O distance in calcite and from the lattice constants in Table 3. The results are summarized in Table 2 and compared to the distances in the corresponding cubic oxides, XO of NaCl type, as well as to the sums of the ionic radii.

For all of the series the oxygen coordination octahedron around the cation is elongate in the direction of c_{hex} as in calcite, the O–O lengths parallel to the basal plane, i.e. to a_{hex}, being less than those inclined to the base. The O–O edge of the regular coordination octahedron in the cubic oxides is intermediate throughout. The elongation of the octahedron in

Table 3. Unit cell dimensions (in Å) of calcite and of other simple rhombohedral carbonates (GRAF, 1961)

		True rhombohedral		Hexagonal		
		a_{rh}	α	a_{hex}	c_{hex}	c_{hex}/a_{hex}
Calcite	$CaCO_3$					
20° C		6.3753	46°4.6'	4.9900	17.061$_5$	3.4191
26° C		6.3760	46°4.3'	4.9899	17.064	3.4197
Rhodochrosite	$MnCO_3$	5.9050	47°43.1$_5$'	4.7771	15.664	3.2790
Siderite	$FeCO_3$	5.7954	47°43.3'	4.6887	15.373	3.2787
Magnesite	$MgCO_3$	5.6752	48°10.9'	4.6330	15.016	3.2411
	$NiCO_3$	5.5795	48°39.7'	4.5975	14.723	3.2024
Smithsonite	$ZnCO_3$	5.6833	48°19.6'	4.6528	15.025	3.2292
Cobaltocalcite	$CoCO_3$	5.6650$_5$	48°33.1'	4.6581	14.958	3.2112
(high-pressure)	$CuCO_3$	5.856	48°11'	4.796	15.48	3.227
Otavite	$CdCO_3$	6.1306	47°19.1'	4.9204	16.298	3.3123

the carbonates obviously reflects something of the charge distribution in
the CO₃ group. At any rate, all O–O distances are larger than two ionic
radii of oxygen. Only for magnesite does the octahedral edge in the basal
plane of 2.93 Å approach the value 2.80 Å after PAULING but is still much
greater than 2.64 Å after GOLDSCHMIDT. Except for magnesite, the cation-oxygen distances are smaller than those in the cubic oxides and
smaller also than the sums of the GOLDSCHMIDT ionic radii. The coincidence in the case of magnesite of the Mg–O length with that of the oxide,
and also with the sum of the radii, is in line with the lower thermal
stability of magnesite in comparison to calcite. This relation perhaps
also explains part of the difficulties which arise in the experimental
preparation of magnesite at ordinary temperatures (see p. 72 ff.).

A substantial part of the above discussion has dealt with the dimensions and the shape of the CO₃ group. The C–O distance as determined
from X-ray diffraction refers to the centers of the electronic systems of C
and O, and these are assumed to be spherically symmetrical in the
evaluation of the diffracted intensities, although deviations from spherical symmetry are allowed for when the anisotropic thermal motion is
studied as in the papers of INKINEN and LAHTI and of CHESSIN et al. At
any rate, the close coincidence of calculated and observed intensities in
these papers, reflected by the rather low R-values (or discrepancy factors)
of about 3% and less, shows that the picture of vibrating spherical atoms
is essentially correct. Since it is not possible to distinguish deviations
from spherical symmetry, if any, from thermal vibrations and since it
would be too difficult in general to show the latter in crystal structure
drawings, individual atoms or ions are usually represented by spheres for
the sake of simplicity. In Figs. 4 and 5a the oxygens of the CO₃ group are

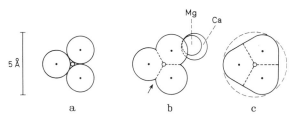

a   b   c

Fig. 5 a–c. Various choices for the graphical representation of CO₃; based on
distances in calcite: 1.28 Å for C–O and 2.22 Å for O–O. a Mutual contact of
oxygen spheres, 2.2 Å in diameter. b Oxygen spheres, 2.6 Å in diameter, to conform with cation-oxygen distance in rhombohedral carbonates and ionic radii,
resulting in mutual overlap of component atoms. The position of the cations, Ca
and Mg, is shown for calcite and magnesite. c Same as b; but by increased merging
of oxygen electrons, the breaks between adjacent oxygens (marked by arrow in b)
have disappeared. The electrons will not extend, however, as far as the broken
circle since it is unlikely that CO₃ approaches axial symmetry

outlined by circles 2.2 Å in diameter, which equals the O–O distance in the CO_3 group in calcite. Thus, CO_3 is thought to be composed of spheres in mutual contact. The near coincidence of the cation-oxygen distances and the sums of ionic radii would call, however, for an oxygen radius of at least 1.30 Å. This would result in an overlap of the oxygens (Fig. 5b) and is in line with a partially covalent C–O bond. CO_3 is thus represented by three bulges. But we cannot be sure whether the notches by which the bulges are now separated have any realistic meaning. The small magnesium which might have been expected to probe the gaps between the oxygens in the structure of magnesite is still farther removed from the gaps than is the calcium in calcite (Fig. 5b). The electrons of adjacent oxygens may also have merged to such an extent that there are no gaps altogether between adjacent oxygens as in Fig. 5c.

This paragraph is not written to add to the speculations concerning the bonding in, and the shape of, the CO_3 group but rather to emphasize that the representation by oxygen spheres in mutual contact, which has been used for its graphical simplicity in Fig. 4 and some of the following illustrations, is highly speculative and merely a convention. Nevertheless, it appears desirable that the shape of CO_3 and its charge distribution be determined more in detail, as far as this is meaningful, by independent structure analyses of more carbonates, among them those listed in Table 3.

2. Superstructures Based on Calcite Pattern

a) Regular Interstratifications

α) Dolomite

The rhombohedral carbonates discussed so far are characterized by one main cationic species. Among the more complex carbonates most of the compounds with two different cations are very similar to calcite and its analogues in their crystallographic properties. The most important example is the rock-forming mineral dolomite, $CaMg(CO_3)_2$. Its lattice dimensions are found in Table 4 along with the data for the other known rhombohedral double carbonates. That dolomite is a double salt and not just a mixed-crystal of calcite type was at first suggested by the narrow range of chemical composition as well as by the morphological symmetry $C_{3i} - \bar{3}$, which is lower than that of calcite $D_{3d} - \bar{3}m$. W. L. BRAGG (1914, p. 488), who carried out the first X-ray diffraction experiments on dolomite tried to explain the structural relation of dolomite to calcite as follows: "It is tempting to consider that in dolomite the arrangement of planes perpendicular to the trigonal axis may be $Ca-CO_3-Mg-CO_3$. This would give the crystal the right symmetry (rhombohedral tetarto-

2*

Table 4. Unit cell dimensions (in Å) of rhombohedral double carbonates

		True rhombohedral		Hexagonal			Symmetry
		a_{rh}	α	a_{hex}	c_{hex}	c_{hex}/a_{hex}	
Dolomite	$CdMg(CO_3)_2$	5.8984	47°47′	4.7770	15.641	3.2742	?
	disordered:	5.9084	47°40′	4.7746	15.678	3.2836	$R\bar{3}$
	$CaMg(CO_3)_2$	6.0154	47°6.6′	4.8079	16.010	3.3299	—
	arithmetic mean of calcite and magnesite	6.0251	47°4.0′	4.8114_5	16.039_5	3.3336	—
	$SrMg(CO_3)_2$ [a]	6.168_6	46°51′	4.905	16.44	3.352	?
	$PbMg(CO_3)_2$ [b]	6.209	46°43′	4.924	16.56	3.363	$R32$
Norsethite	$BaMg(CO_3)_2$ [c]	6.296	46°58′	5.017	16.77	3.343	$R32$
Kutnohorite	$CaMn(CO_3)_2$ disordered:			4.85	16.34	3.36_9	$R\bar{3}$?
	$Ca_{1.03}Mn_{0.84}Mg_{0.115}Fe_{0.015}(CO_3)_2$ [d]	6.1401	46°50′	4.8797	16.367	3.3541	—
Ankerite	$CaFe(CO_3)_2$			4.819	16.10	3.341	$R\bar{3}$?
	$Ca_{1.03}Mg_{0.63}Fe_{0.33}(CO_3)_2$ [e]	6.0855	46°51.5′	4.8393	16.218_5	3.3514	—
	arithmetic mean of calcite and siderite						

Source: GRAF (1961) unless otherwise stated.
For the ordered minerals kutnohorite and ankerite, values for ideally composed materials are not available. The lattice dimensions here quoted apply to the formulae [d] and [e].
[a] FROESE (1967); [b] LIPPMANN (1966); [c] LIPPMANN (1968a); [d] FRONDEL and BAUER (1955); [e] HOWIE and BROADHURST (1958).

hedral), the digonal axes being destroyed by the substitution of magnesium atoms for half of the calcium atoms."

This prediction has been confirmed by later investigators. WASA-
STJERNA (1924) and WYCKOFF and MERWIN (1924) assigned dolomite to
the space group $C_{3i}^2 - R\bar{3}$ on the basis of observed and extinguished X-ray
reflections and of diffraction symmetry. The latter has been found to
be lower than in calcite, in accordance with the morphological symmetry
of the crystals. The authors also found that for the proper description of
the space symmetry of dolomite the cleavage rhombohedron is not useful on the same grounds as was the case for calcite. An elongate rhom-

Table 5a. Atomic sites in dolomite structure. Space group $R\bar{3} - C_{3i}^2$ (No. 148;
International Tables I)
Occupancy of rhombohedral cell

	Wyckoff notation	Point symmetry	Coordinates
Ca	a	$\bar{3}$	$0, 0, 0.$
Mg	b	$\bar{3}$	$\frac{1}{2}, \frac{1}{2}, \frac{1}{2}.$
2 C	c	3	$x_C, x_C, x_C;$ $-x_C, -x_C, -x_C.$
6 O	f	1	$x, y, z;$ $z, x, y;$ $y, z, x;$
			$-x, -y, -z;$ $-z, -x, -y;$ $-y, -z, -x.$

The coordinates are fractions of a_{rh}. They are related to the hexagonal coordinates $x_{hex}, y_{hex}, z_{hex}$ in Table 5b by the transformation:

$$x_{rh} = \quad x_{hex} \quad\quad + z_{hex}$$
$$y_{rh} = -x_{hex} + y_{hex} + z_{hex}$$
$$z_{rh} = \quad\quad - y_{hex} + z_{hex}$$

Table 5b. Occupancy of hexagonal unit cell

	Wyckoff notation	Point symmetry	Coordinates
3 Ca	a	$\bar{3}$	$0, 0, 0.$
3 Mg	b	$\bar{3}$	$0, 0, \frac{1}{2}.$
6 C	c	3	$0, 0, z_C;$ $0, 0, -z_C.$
18 O	f	1	$x, y, z;$ $-y, x-y, z;$ $y-x, -x, z;$
			$-x, -y, -z;$ $y, y-x, -z;$ $x-y, x, -z.$
			$+(0, 0, 0;$ $\frac{1}{3}, \frac{2}{3}, \frac{2}{3};$ $\frac{2}{3}, \frac{1}{3}, \frac{1}{3})$

Cf. legend to Table 1b. The hexagonal coordinates of this table are related to
the rhombohedral ones in Table 5a by the transformation

$$x_{hex} = \tfrac{2}{3}x_{rh} - \tfrac{1}{3}y_{rh} - \tfrac{1}{3}z_{rh}$$
$$y_{hex} = \tfrac{1}{3}x_{rh} + \tfrac{1}{3}y_{rh} - \tfrac{2}{3}z_{rh}$$
$$z_{hex} = \tfrac{1}{3}x_{rh} + \tfrac{1}{3}y_{rh} + \tfrac{1}{3}z_{rh}.$$

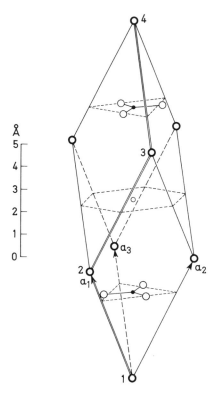

Fig. 6. Rhombohedral unit cell of dolomite, containing one formula $CaMg(CO_3)_2$. Open circles denote centers of Ca, O, and Mg in order of decreasing size. Small closed circles: C. CO_3 is thus represented by threefold "dumb-bells". The calcium atoms marked 1, 2, 3, 4 serve to show the position of the rhombohedral cell with respect to the hexagonal cell in Fig. 7

bohedron, of nearly the same shape as in calcite and containing one formula $CaMg(CO_3)_2$, is the true unit cell. This cell is presented in Fig. 6, and the atomic coordinates available for the dolomite composition are given in Table 5a. As in the case of calcite, we may expect a more coherent picture of the structure from the hexagonal description in Table 5b (Fig. 7).

By making $z_C = z_O = \frac{1}{4}$ and $y = 0$, WASASTJERNA (1924) proposed a simplified model for dolomite, in which CO_3 has the same coordinates as in calcite (Table 1b). However, two different cations, Ca and Mg, occur and these are segregated in alternate layers as suggested by W. L. BRAGG (1914). From the point of view of the cation distribution dolomite is a superstructure based on the calcite arrangement.

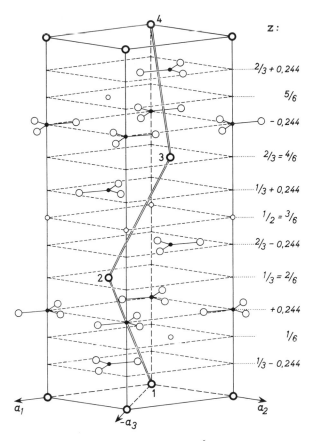

z:

2/3 + 0,244

5/6

- 0,244

2/3 = 4/6

1/3 + 0,244

1/2 = 3/6

2/3 - 0,244

1/3 = 2/6

+ 0,244

1/6

1/3 - 0,244

Fig. 7. Hexagonal unit cell of dolomite, height 16.01 Å, containing $3\,CaMg(CO_3)_2$. The components are represented the same way as in Fig. 6. The z coordinates, in fractions of c, apply to the cation and CO_3 levels. The calcium atoms (ions) marked 1, 2, 3, 4 show orientation of rhombohedral cell (Fig. 6); see also Fig. 3

WASASTJERNA's model would yield the same value for the Ca–O and Mg–O distances. This is in conflict with the bond lengths in calcite and magnesite. The model is thus an approximation. To conform with the cation–CO_3 distances in the simple carbonates, we may try to build up the dolomite structure from the ultimate layers of magnesite and calcite. Such a layer of calcite has been shown in Fig. 4. Its thickness, determined by the difference in z coordinates of the CO_3 levels above and below Ca, is $\frac{c}{6}$. In this way a rhombohedral carbonate layer can be thought to consist of a cation level and the adjacent halves of the CO_3 layers above and below. To accommodate an alternation of such magnesite and cal-

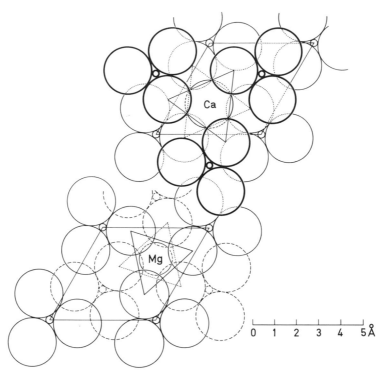

Fig. 8. Individual calcium carbonate layer and magnesium carbonate layer of dolomite structure projected in the direction of the hexagonal c axis, showing lozenge-shaped outlines of hexagonal unit cells. The CO_3 level (heavy circles) above the Ca has $z = \frac{2}{3} - 0.244$; Ca: $z = \frac{1}{3}$. The CO_3 level below (here dotted) the Ca continues (full circles) on top of the Mg ($z = \frac{1}{6}$) and has $z = 0.244$. The CO_3 level (broken circles) below the Mg has $z = \frac{1}{3} - 0.244$ (see Fig. 7). The coordination octahedra around Ca and Mg are shown by their basal equilateral triangles (cf. Fig. 4). They are not regular octahedra proper, but rather combinations of a rhombohedron and basal planes

cite layers within the c period of dolomite, the parameter $z_C = z_O$ has to be less than the value in the simplified model of $\frac{1}{4} = 0.25$. As a result, the oxygens are closer to magnesium than to calcium, as seen in Fig. 7, in accordance with the thickness relation of both component layers. Because magnesite and calcite have different a_{hex} dimensions, the CO_3 groups of the layers taken from the simple carbonates will not match unless the pattern undergoes some distortion. This can be effected by rotating the CO_3 groups, from their original alignment parallel to the a_{hex} directions, in such a sense as to approach the oxygens to magnesium. Matching calcium and magnesium carbonate layers, linked by a common CO_3 level, are shown in Fig. 8. That these layers are similar but no longer identical to those in the simple carbonates (LIPPMANN and JOHNS, 1969), is seen by comparing Figs. 4 and 8.

In order to comply with the dimensional requirements of the simple carbonates, the special coordinates, $z = \frac{1}{4}$ and $y = 0$, assigned by WASA-STJERNA have to be modified. This means that the full variability of the general oxygen parameters x, y, z in Table 5b is needed to depict the actual structure of dolomite. The simplified model has been useful, however, in seeing the superstructure relation to calcite more clearly.

An approximate determination of the oxygen parameters in dolomite has been carried out by BRADLEY, BURST and GRAF (1953). They used plausible interatomic distances as a starting point and arrived at a model which yielded a fair agreement of calculated and observed X-ray diffraction intensities, as measured by the powder method. STEINFINK and SANS (1959) have published a refinement of this model for which they used 501 reflections from a single crystal. In their calculations they tried different z coordinates for carbon (z_C) and oxygen (z_O), i.e. they allowed for a possible pyramidal deformation of the CO_3 group. This would be compatible with the space group of dolomite, whereas in calcite the planarity of CO_3 is secured by symmetry. It turned out that the parameters $z_C = 0.2435 \pm 0.00031$ and $z_O = 0.2440 \pm 0.00017$ agree within the limits of accuracy. The flat configuration of CO_3 applies thus also to dolomite.

The lattice dimensions, $a_{hex} = 4.815$; $c_{hex} = 16.119$ Å, which STEINFINK and SANS give in their paper, appear high compared to those of GRAF (1961) in Table 4. Since the authors did not state the method by which they obtained their values, it was deemed preferable to use only their positional parameters in the calculation of the interatomic distances, but to use the values of GRAF (1961) for lattice constants. These incidentally agree closely with the data of HOWIE and BROADHURST (1958): $a_{hex} = 4.810$; $c_{hex} = 16.02$ Å.

The interatomic distances thus obtained for dolomite are listed in Table 6 and compared to the values of the simple carbonates, magnesite and calcite. The C–O length of 1.281 Å is identical with the value for calcite based on INKINEN and LAHTI. As in the simple carbonates, the coordination octahedra around the cations Mg and Ca are both elongate in the direction of the c axis, as shown by the shorter O–O distance parallel to the basal plane in comparison to the inclined O–O length. Mg–O is shorter than in magnesite and Ca–O is longer than in calcite. In order to elucidate this slight though distinct discrepancy, a model for dolomite was constructed with the same cation-oxygen distances as in the simple carbonates. The result is shown by the values listed under "model" in Table 6. It is noteworthy that the coordination octahedra

Table 6. Interatomic distances (in Å) in dolomite and related carbonates

		Mg–O octahedron			Ca–O octahedron		
In basal plane	C–O	1.281 (throughout)					
		ST. and S.	magnesite	model	ST. and S.	calcite	model
	O–O	2.911	2.930	2.941	3.299	3.263	3.262
Out of basal plane	O–O	2.993	3.021	3.013	3.436	3.411	3.414
	X–O	2.088	2.104	2.105	2.382	2.360	2.361

ST. and S.: Parameters of STEINFINK and SANS (1959); corrected (GRAF, 1969): $x = 0.2474$; $y = -0.0347$; $z = 0.2440$; lattice constants: $a = 4.8079$; $c = 16.010$ (GRAF, 1961).

Magnesite and calcite: C–O distance based on parameter for calcite of INKINEN and LAHTI (1964) and lattice constants of GRAF (1961), as in Table 2.

Model: calculated for dolomite to conform to cation-oxygen distances of calcite and magnesite; parameters: $x = 0.2509$; $y = -0.0287$; $z = 0.2444$. Note that these values are very close to those proposed by BRADLEY, BURST and GRAF (1953): $x = 0.257$; $y = -0.028$; $z = 0.243$.

around the cations are elongate also in this model and obtain very nearly the same dimensions as in the simple carbonates. It may be a coincidence that the parameters x and y of the model (see legend to Table 6) are also near those proposed by BRADLEY et al. (1953). However, in the light of this situation, one would like to see the dolomite structure reexamined and refined to the extent that has been done for calcite. If the deviation of the distances of dolomite from those of the simple carbonates turns out to be real and can be explained in terms of bonding and crystal energy, this might throw new light on dolomite as a superstructure.

For the moment, it may be concluded from the structural analogy with calcite that the bonding between cations and CO_3 in dolomite is essentially ionic. In particular, the cleavage along the faces of the morphological reference rhombohedron is then explained the same way as in calcite and NaCl. In classical crystallography the cleavage rhombohedron of both calcite and dolomite has been indexed (100) with respect to a rhombohedral coordinate system, whose axes are parallel to the cleavage edges. For the same reasons that a hexagonal reference system is preferred in structural crystallography, the cleavage rhombohedron has also been assigned the hexagonal indices (10$\bar{1}$1). When the orientation of morphological features, such as cleavage, are to be viewed with respect to crystal structure, it is convient to follow the conventions of X-ray crystallography and to base the indices on one of the unit cells. The cleavage rhombohedron is then indexed either (211) with respect to the true rhombohedral cell, or (10$\bar{1}$4) in the hexagonal description. Both

Table 7. Correlation of indices for rhombohedral simple and double carbonates for various reference systems (after BRADLEY, BURST and GRAF, 1953)

Rhombohedral unit cell (X-ray)[a]	Cleavage rhombohedron (morphologic)[b]	Hexagonal choice based on cleavage rhombohedron (morphologic)[b]	Hexagonal structure cell (X-ray)[a]
<u>111</u>	111	0001	<u>0003</u>
<u>100</u>	31̄1̄	4̄041	<u>101̄1</u>
110	11̄1̄	02̄2̄1	011̄2
211	100	101̄1	101̄4
222	111	0001	0006
<u>221</u>	33̄1̄	04̄4̄5	<u>011̄5</u>
101̄	101̄	11̄2̄0	11̄20
210	51̄3̄	44̄8̄3	112̄3
<u>11̄1</u>	33̄5̄	08̄8̄1	<u>02̄21</u>
200	31̄1̄	8̄081	202̄2
<u>322</u>	511	4̄047	<u>101̄7</u>
220	11̄1̄	02̄2̄1	022̄4
332	110	011̄2	011̄8
321	31̄1̄	22̄4̄3	112̄6
<u>333</u>	111	0001	<u>0009</u>
<u>311</u>	71̄1̄	8̄085	<u>202̄5</u>

[a] Superstructure reflections are underlined.

[b] Common factors, fractional or integer, have been eliminated from the morphological indices. Therefore, identical morphological indices are written for various orders of X-ray indices. Cf. CORRENS, pp. 350—351.

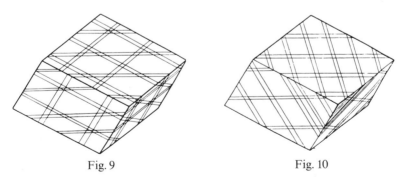

Fig. 9   Fig. 10

Fig. 9. Cleavage rhombohedron (morphological reference rhombohedron) of calcite showing pressure twin lamellae (from CORRENS), according to the morphological indices: (011̄2) (hex) or (110) (rh); X-ray indices: (011̄8) (hex) or (332) (rh)

Fig. 10. Orientation of pressure twin lamellae in dolomite (from CORRENS), described by the morphological indices: (02̄21) (hex) or (11̄1) (rh); X-ray indices (011̄2) (hex) or (110) (rh)

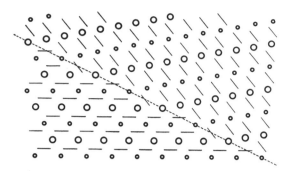

Fig. 11. Lower left: (11.0) projection of dolomite structure (cf. Fig. 7), separated by alleged twin boundary (dashed) from deformed part (upper right). The mutual orientation is such as would occur if the calcite type twinning parallel to $(01\bar{1}2)_{morph.}$ or $(01\bar{1}8)_{X\text{-}ray}$ were produceable by pressure also in dolomite. However, the deformed part no longer exhibits the segregation of the cations Ca (large circles) and Mg (small circles) in alternating levels parallel to the CO_3 groups (lines) which characterizes the original crystal (lower left). This type of deformation thus fails to reproduce the dolomite pattern on both sides of the boundary. If it would indeed occur, a new polymorph of dolomite would be synthesized in the deformed part, for which there is no observational evidence. – If the two differently represented cations are assumed to be identical, we have the structure of calcite. The arrangements on both sides of the twin boundary are then identical and in symmetrical positions with respect to the latter. Such a pattern may be produced by a succession of small uniform displacements of the cations parallel to the twin boundary. The CO_3 groups are thought to be rotated to symmetrical positions by the cations. In this way, by presuming the orientation relations actually observed in pressure twinned calcite a reasonable picture is obtained of the atomic arrangement near the twin boundary

choices are referred to as "X-ray indices". A correlation for the indices of other prominent faces is given in Table 7.

It is interesting to note that dolomite exhibits an orientation of pressure twin lamellae which is different from calcite (compare Figs. 9 and 10), although both structures are similar. Nevertheless, according to BRADLEY, BURST and GRAF (1953), both types of characteristic orientations may be deduced from the respective crystal structures. Interpretations following these authors are outlined in the legends to Figs. 11 and 12. They imply charged components held together by ionic forces. By virtue of their non-directional character these forces allow the reorientations accompanying the twin glide, and, at the same time, secure the cohesion of the deformed part of a crystal and its coherence with the undeformed part along the twin boundary.

It has been shown above that dolomite differs from calcite and the other rhombohedral carbonates by its lower symmetry, which is due to

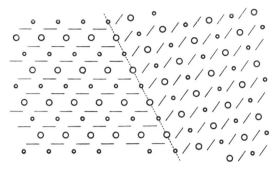

Fig. 12. (11.0) projection of dolomite structure (cf. Fig. 7), showing twin boundary (dashed) parallel $(02\bar{2}1)_{morph.}$ or $(01\bar{1}2)_{X-ray}$, as may be produced by pressure. This type of deformation results in identical and symmetrical arrangements on both sides of the boundary. It is noteworthy that the deformed part exhibits a regular alternation of Ca (large circles) and Mg (small circles) levels parallel to the planes of CO_3 (lines) as does the undeformed part. The rotation of the CO_3 is of smaller extent than in Fig. 11, but it may be noted that the break in a given lattice row is greater than there. Therefore, this type of pressure twinning does not occur in calcite where the stable configuration is attained at smaller displacements parallel to the twin boundary. – Figs. 11 and 12 are from BRADLEY, BURST and GRAF (1953) (reprinted with permission from *The American Mineralogist*) whose paper may be consulted for more details

the presence of two different cations segregated in alternating layers. However, dolomite-like minerals which deviate from this ideal ordering are being found as constituents of sediments and sedimentary rocks with increasing frequency. They will be discussed later on pp. 47—50. Although pertinent investigations so far published do no warrant an estimate of overall abundances, it appears that regardless of the recent findings of disordered dolomite-like minerals, the importance of ideally ordered dolomite as a constituent of sedimentary rocks remains unchallenged.

Among the determinative methods applicable to rock-forming minerals, only X-ray diffraction affords a straightforward evaluation of symmetry and order in dolomite. The X-ray study of both these aspects is possible only in single crystals, whereas powder diffraction is sufficiently sensitive for a cursory appraisal of order. Since the segregation of the cations is the most salient feature of dolomite, it has also become an important determinative criterion. Cation order is indicated in X-ray powder patterns by the presence of reflections which occur in addition to those observed in the simple rhombohedral carbonates. These additional reflections may be discerned by space-group extinctions. In the space group $R\bar{3} - C_{3i}^2$ of dolomite, reflections indexed hkl on the basis of the true rhombohedral unit cell occur for all possible combinations when h, k and l assume non-fractional numbers including zero, i.e., there are no systematic extinctions. When the reflections of dolomite are transformed to indices referring to the hexagonal unit cell, we obtain only such combinations $hk.l$ for which the sum $-h+k+l$ is a multiple of three. Reflections with index combinations not satisfying this condition are said to be extinguished for the hexagonal lattice. These extinctions follow from the fact that in the hexagonal unit cell certain aspects of the structure such as the layers perpendicular to c, are repeated three times, and they apply to both calcite and dolomite. In the latter, by the different occupancy of the cation levels on either side of a CO_3 layer, a given oxygen is bonded by two different cations. In calcite the two calcium ions coordinated with an oxygen are related by a binary axis through the oxygen and the carbon. The cooperation of these axes with the centers of symmetry in calcium is equivalent to glide planes parallel to the c-axis. X-ray reflections that would come from certain planes perpendicular to a glide plane are then extinguished. Their indices are of the type hhl, with odd-numbered l, for the rhombohedral description. For the hexagonal choice, in addition to rhombohedral extinctions stated above, reflections indexed $h0.l$ and $0k.l$, with odd-numbered l, are absent in calcite. Since they do occur in dolomite, which may be thought to be derived as a superstructure from calcite, these reflections are generally referred to in carbonate mineralogy as "superstructure reflections". Their appearance in dolomite, contrasted with their absence in calcite, is illustrated in Fig. 13. Their intensities may be used as an empirical index of order. However, it must be cautioned that the intensity of the superstructure lines may also be attenuated by preferred orientation with respect to the rhombohedral cleavage faces of the crystallites. Measures must be taken, therefore, to minimize preferred orientation in the powder to be X-rayed. Prolonged grinding has to be avoided because it tends to destroy the cation order in dolomite, as shown by BRADLEY, BURST and GRAF (1953).

β) Norsethite

Besides dolomite, norsethite, $BaMg(CO_3)_2$, is a rhombohedral double carbonate for which the structure is known in some detail. Although norsethite was first described from a sedimentary environment, the sodium-carbonate bearing Green River formation of Wyoming (MROSE, CHAO, FAHEY and MILTON, 1961), it is too rare to figure as a rock-forming mineral. Only three other occurrences have since been reported, and these are in non-sedimentary formations (KAPUSTIN, 1965; SUNDIUS and BLIX, 1965; STEYN and WATSON, 1967). Nevertheless, norsethite is of some interest to sedimentary petrology on account of the ease of its formation at room temperature under suitable experimental condi-

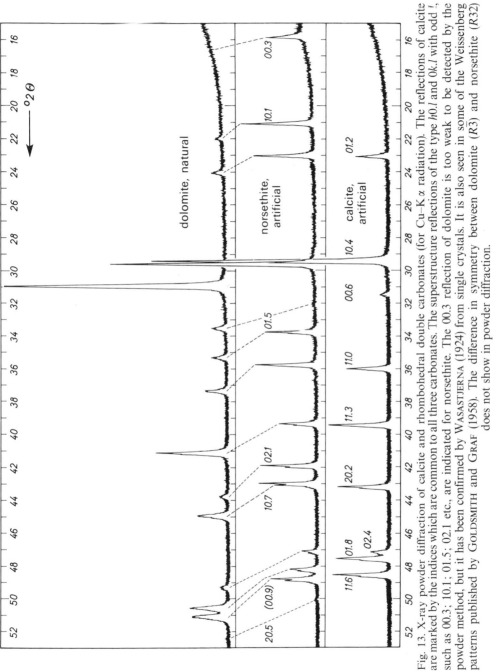

Fig. 13. X-ray powder diffraction of calcite and rhombohedral double carbonates (for Cu-Kα radiation). The reflections of calcite are marked by the indices which are common to all three carbonates. The superstructure reflections of the type h0.l and 0k.l with odd l, such as 00.3; 10.1; 01.5; 02.1 etc., are indicated for norsethite. The 00.3 reflection of dolomite is too weak to be detected by the powder method, but it has been confirmed by WASASTJERNA (1924) from single crystals. It is also seen in some of the Weissenberg patterns published by GOLDSMITH and GRAF (1958). The difference in symmetry between dolomite (R3̄) and norsethite (R32) does not show in powder diffraction.

tions. From these, conclusions regarding the normal-temperature formation of dolomite are possible, which will be discussed on pp. 178—190. It will be seen that the most important aspect justifying such conclusions from analogy is the occurrence of magnesite-like layers in the crystal structures of both dolomite and norsethite (p. 168). It is thus appropriate to include the norsethite structure in this monograph and to describe it with special reference to dolomite.

Regardless of the complete analogy of lattice dimensions (Table 4) and X-ray powder patterns (Fig. 13), both minerals cannot be isostructural, because, with respect to dolomite, norsethite exhibits a higher symmetry in single-crystal X-ray diffraction (MROSE et al., 1961). However, the presence of superstructure reflections of the same type as in dolomite, which are seen in Fig. 13, suggests that the two different cations of norsethite are ordered in alternating basal levels. Such an arrangement which is compatible with the required higher diffraction symmetry may be found in the space group $R32 - D_3^7$. The atomic sites available for the norsethite composition are listed in Tables 8a and 8b. Numerical values for the coordinates of the oxygens sites were deter-

Table 8a. Atomic sites in norsethite structure. Space group $R32 - D_3^7$ (No. 155 International Tables I)
Occupancy of rhombohedral unit cell

	Wyckoff notation	Point symmetry	Coordinates		
Ba	a	32	$0, 0, 0.$		
Mg	b	32	$\frac{1}{2}, \frac{1}{2}, \frac{1}{2}.$		
2 C	c	3	$x_C, x_C, x_C;$	$-x_C, -x_C, -x_C.$	
6 O	f	1	$x, y, z;$	$z, x, y;$	$y, z, x;$
			$-y, -x, -z;$	$-z, -y, -x;$	$-x, -z, -y.$

See also explanation under Table 5a.

Table 8b. Occupancy of hexagonal unit cell

	Wyckoff notation	Point symmetry	Coordinates		
3 Ba	a	32	$0, 0, 0.$		
3 Mg	b	32	$0, 0, \frac{1}{2}.$		
6 C	c	3	$0, 0, z_C;$	$0, 0, -z_C.$	
18 O	f	1	$x, y, \quad z;$	$-y, x-y, \quad z;$	$y-x, -x, \quad z;$
			$y, x, -z;$	$-x, y-x, -z;$	$x-y, -y, -z.$
				$+ (0, 0, 0;$	$\frac{1}{3}, \frac{2}{3}, \frac{2}{3}; \quad \frac{2}{3}, \frac{1}{3}, \frac{1}{3})$

See explanations under Tables 1b and 5b.

Table 9. Oxygen coordinates and interatomic distances (in Å) for norsethite and PbMg(CO₃)₂ (LIPPMANN, 1968a)

XMg(CO₃)₂			BaMg(CO₃)₂	PbMg(CO₃)₂
Oxygen coordinates		x	0.199	0.214_5
as fractions of a_1; a_2		$-y$	0.089	0.075
and c(hex)		$z(=z_C)$	0.242	0.245
In basal plane:				
CO₃ group	{	C–O	3× 1.281	1.281
	{	O–O	3× } 2.219	2.219
X–O polyhedron			6× }	
Mg–O octahedron --->		O–O	6× 2.89	2.85
Out of basal plane:				
	{	X–O (short)	6× 2.71_5	2.58
X–O polyhedron	{	X–O (long)	6× 3.18	3.17
	{	O–O	6× 3.09	2.95
	{	Mg–O	6× 2.09	2.09
Mg–O octahedron	{	O–O (short)	3× 2.78	2.78
	{	O–O (long)	3× 3.30	3.40

The coordinates for PbMg(CO₃)₂ are based on an electrostatic model with approximate maximum stability.

mined by LIPPMANN (1968a) from X-ray powder diffraction intensities. They are found in Table 9 along with the resulting interatomic distances. Corresponding values for PbMg(CO₃)₂ are included. Figs. 14 and 15 show the rhombohedral and hexagonal unit cells, respectively. Ba and Mg occupy the same positions as the cations in dolomite and thus form an array of regularly alternating layers oriented perpendicular to the c axis, as predicted from the presence of the superstructure reflections.

Norsethite differs from dolomite in the articulation of the CO₃ groups. It is true that the oxygen parameters x, y, z are of similar magnitudes in both structures, but they are dissimilar in proportion.

In dolomite the C–O directions are rotated by 6.5° away from the projections of the a_{hex} axes (Fig. 8), whereas the corresponding angular distance in norsethite is 17.5° (Fig. 16). There is a perceptible difference in CO₃ orientation already in the layers with z, $\frac{1}{3}+z$ and $\frac{2}{3}+z$, and the CO₃ alignment for $-z$, $\frac{1}{3}-z$, and $\frac{2}{3}-z$ follows a fundamentally different pattern. The latter set of layers is related to the former by centers of symmetry in the cations in dolomite and by binary axes through the cations in norsethite. Nevertheless, magnesium is coordinated by six oxygens forming octahedra in both minerals. The different symmetry of norsethite

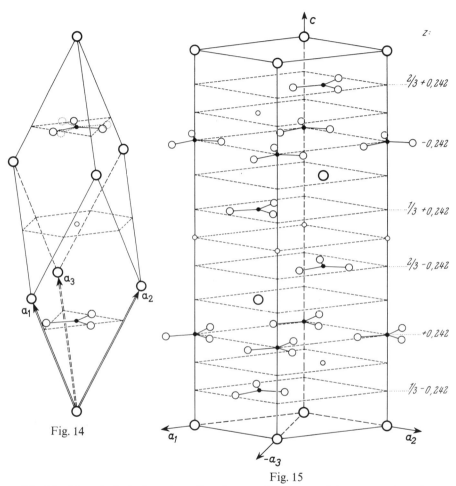

Fig. 14

Fig. 15

Fig. 14. Rhombohedral unit cell of norsethite, containing one formula $BaMg(CO_3)_2$. Ba, O and Mg are represented by open circles in order of decreasing size. Small closed circles: C. The upper CO_3 group drawn in full lines is related to the lower by binary axes in Mg. The dotted CO_3 group shows the orientation which would be generated from the lower if a center of symmetry were operative in Mg as in dolomite (see Fig. 6)

Fig. 15. Hexagonal unit cell of norsethite, height $c = 16.77$ Å, containing $3 BaMg(CO_3)_2$. The components are denoted as in Fig. 14. The z coordinates, in fractions of c, refer to the CO_3 levels.
The cations have the same z coordinates as in Fig. 7

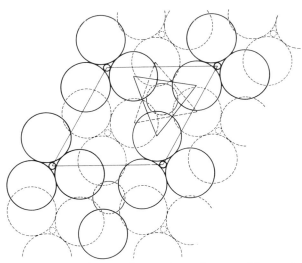

Fig. 16. Individual magnesium carbonate layer in norsethite projected in the direction of the hexagonal c axis. Upper CO_3 level (full lines): $z = 0.242$, Mg (partly broken lines): $z = \frac{1}{6}$, Lower CO_3 (broken lines): $z = \frac{1}{3} - 0.242$. – The oxygen octahedron around Mg is described by its upper (full) and lower (broken) basal triangles. It is distorted with regard to a regular octahedron by a mutual twist of these triangles. It is actually a trigonal trapezohedron truncated at the corners by two basal planes.
Edge length of lozenge-shaped outline of hexagonal unit cell: $a_{hex} = 5.017 \, \text{Å}$

correlates with a greater degree of distortion of the oxygen octahedron surrounding the magnesium. In the absence of symmetry centers in the cation, the basal oxygen triangles are twisted with respect to each other under the influence of the binary axes (Fig. 16). Consequently, the octahedron has three different O–O edge lengths, although the Mg–O distance does not differ significantly from that in dolomite.

In view of the size difference between barium and calcium, it is not surprising that the bonding of the large cation constitutes the main distinguishing feature. In contrast with the octahedral sixfold coordination of the calcium in dolomite, the barium is surrounded by 12 oxygens. The coordination polyhedron is described by two basal hexagons which are shown in Fig. 17. It has almost the shape of a ditrigonal prism in which three of the oxygens in either base are closer to the barium than the other three, resulting in two different Ba–O lengths for two sets of six oxygens. The short Ba–O length of 2.715 Å is slightly shorter than the sum of the ionic radii of 2.75 Å. According to the symmetry $R32$ the oxygen prism is distorted in a trapezohedral fashion. It is shown in perspective in Fig. 18.

3*

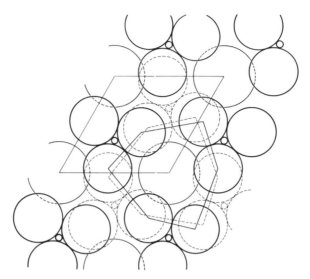

Fig. 17. Individual barium carbonate layer in norsethite projected in the direction of the hexagonal c axis. Upper CO_3 level (full lines): $z = \frac{1}{3} - 0.242$, Ba (partly broken lines): $z = 0$. Lower CO_3 level (broken lines): $z = 0.242 - \frac{1}{3}$. – The outlines of the oxygen coordination polyhedron are shown around one of the bariums. The outline of the hexagonal unit cell is marked as in Fig. 16

We may summarize the structure of norsethite as a regular interstratification of two types of component layers like that of dolomite (cf. LIPPMANN and JOHNS, 1969). Dolomite has been described as an alternation of magnesium and calcium carbonate layers which are similar to those occurring in calcite and magnesite. In norsethite, only the magnesium carbonate layer (Fig. 16) may be thought of as derived from a simple mineral, and we may speak of a distorted magnesite layer. The other component of the interstratification, the barium carbonate layer (Fig. 17), bears no resemblance to any known structure of a simple carbonate, and it is perhaps unique to norsethite and its isotypes.

PbMg$(CO_3)_2$ is isostructural with norsethite by virtue of its crystal symmetry (LIPPMANN, 1966). The only crystallographic data available for SrMg$(CO_3)_2$ are its X-ray powder reflections (FROESE, 1967). In view of the similar sizes of Pb^{2+} and Sr^{2+}, it is likely to be another isotype of norsethite. In this case, the bonding of the large cation in norsethite will give us an idea of how cations as large as strontium may be accommodated as trace elements in the structures of rhombohedral carbonates such as dolomite and calcite.

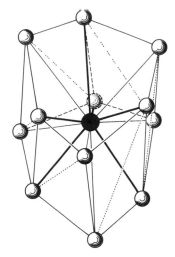

Fig. 18. Norsethite. Coordination polyhedron around barium (closed sphere), formed by twelve oxygens (shaded spheres) arranged in two basal planes. Two oxygens connected by the short basal edge belong to one CO_3 group. The lateral faces are quadrangles only by approximation. They are slightly bent outward along the dotted diagonals. From the crystallographic point of view, the coordination polyhedron is made up of a combination of two steep trapezohedra with slightly different orientations and two basal planes. These latter are symmetrical to the central Ba, and truncate the trapezohedra in their zig-zag belt. The short Ba–O bond is marked by heavier lines than the long one

b) Superstructures with Order in the Individual Cation Layer

If we think of a single basal cation layer of a calcite-type structure, it is conceivable that different species are accommodated in an ordered fashion. Such superstructures will be characterized by a_{hex} dimensions which are of greater magnitude than the a_{hex} of the rhombohedral carbonates discussed so far. However, the intercation distances within such a layer will be similar in magnitude to the a_{hex} values known from the simple carbonates. Among the variety of such patterns ("tessellations") which are theoretically possible in a hexagonal network (see TAKEDA and DONNAY, 1965), two have been encountered which define complex rhombohedral carbonates. Huntite, $Mg_3Ca(CO_3)_4$, exhibits an a_{hex} edge which is doubled with respect to the mesh a' of the underlying calcite-type network (Fig. 19a). Benstonite contains 13 cations in its fundamental lozenge, whose edge, a_{hex}, is related to that of a calcite-type subcell, a', by $a_{hex} = a' \cdot \sqrt{13}$. Its cation distribution, for which the general frame is shown in Fig. 19b, is not yet known in detail, but the chemical analysis suggests either $Ca_7Ba_6(CO_3)_{13}$ or $MgCa_6Ba_6(CO_3)_{13}$ as possible formulae.

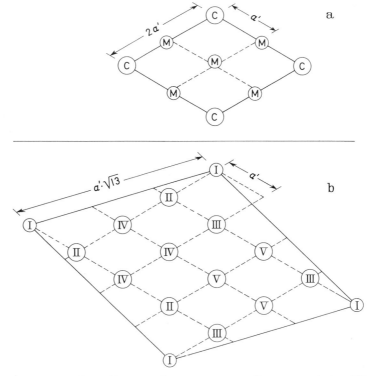

Fig. 19 a and b. Outlines of hexagonal superstructure cells accommodating different species in ordered calcite-type cation layers. The calcite sub-cells, characterized by the dimension a', are shown by broken lines. a The four cations of huntite, one Ca ('C') and three Mg ('M'), require doubled a_{hex} periodicities. The Mg are slightly shifted from the positions of an ideal calcite-type structure (GRAF and BRADLEY, 1962). b The fundamental lozenge of benstonite accommodates 13 cations. Sites denoted by equal Roman numerals are occupied by the same species. The unique site I should contain the Mg of the formula $MgCa_6Ba_6(CO_3)_{13}$. For the definite assignment of the other cations a detailed crystal structure analysis has to be carried out (LIPPMANN, 1962)

Benstonite is a rare mineral which has been reported from a limited number of mineral deposits. It has been mentioned here to give a broader view of a type of superstructure which is based on a principle other than regular interstratification and of which huntite is a representative of some consequence. It is possible that other carbonate superstructures of this type exist and have not so far been recognized.

Huntite was first described by FAUST (1953) from carbonate-bearing volcanic tuffs at Currant Creek, Nevada. It seems to have formed as a

late product of the reaction of meteoric waters with magnesite and dolomite. The latter minerals are supposed by the author to be of hydrothermal origin. More occurrences of huntite have since been reported.
They have been reviewed recently by KINSMAN (1967). It is a common
characteristic of all these occurrences that huntite is associated in some
way or other with dolomite and/or magnesite, from which minerals it is
assumed to have originated by alteration under the influence of near-
surface waters. KINSMAN (1967) himself found huntite on the Trucial
Coast in the Persian Gulf. It occurs there in sub-recent supratidal calcareous sediments which show incipient dolomitization. The rôle of huntite
is thus not limited to being an alteration product. According to KINS-
MAN it may be regarded as a precursor of dolomite and so is of more
general petrological importance.

Table 10. Hexagonal unit cell dimensions (in Å) of huntite and benstonite, analyzed
for superstructure relations and compared to the data of simple carbonates

	Huntite	Magnesite	Calcite	Benstonite
a_{hex}:	$9.505 = 2 \cdot 4.752$	4.633	4.990	$5.07 \cdot \sqrt{13} = 18.28$
c_{hex}:	$7.821 = \dfrac{15.642}{2}$	15.016	17.061	$\dfrac{17.34}{2} = 8.67$

So far, huntite has been found exclusively as polycristalline masses of
submicroscopic grain size. In the absence of suitable single crystals,
GRAF and BRADLEY (1962) have analyzed the structure on the basis of
the X-ray powder pattern. It was indexed by the authors on the basis of
a rhombohedral unit cell of the approximate shape of the calcite cleavage rhombohedron, which contains one formula $Mg_3Ca(CO_3)_4$ and measures $a_{rh} = 6.075$Å with $\alpha = 102° 56'$. The hexagonal unit cell containing
3 formulae has the dimensions $a_{hex} = 9.505$ Å and $c_{hex} = 7.821$ Å. c_{hex} is
about halved with regard to the c_{hex} values known for the simple rhombohedral carbonates, and a_{hex} is doubled. This doubling is implied by the
cation distribution shown in Fig. 19a. The dimensional relationships of
huntite and also of benstonite to the structures of the simple calcite-type
carbonates are demonstrated in Table 10. By adopting the space group
$R32$, GRAF and BRADLEY deduced a distribution of the huntite composition which is accommodated in a cell of the above quoted dimensions.
The atomic sites of this model and their coordinates are quoted in
Tables 11a and 11b.

Table 11a. Atomic sites in huntite structure after GRAF and BRADLEY (1962).
Space group $R32 - D_3^7$ (No. 155; International Tables I)
Occupancy of rhombohedral unit cell

	Wyckoff notation	Point symmetry	Coordinates
Ca	a	32	$0, 0, 0.$
C_I	b	32	$\frac{1}{2}, \frac{1}{2}, \frac{1}{2}.$
(basal plane carbonate group)			
3 Mg	d	2	$0, x, -x;\quad -x, 0, x;\quad x, -x, 0.$
			with $x = 0.541$
3 C_{II}	e	2	$\frac{1}{2}, x, -x;\quad -x, \frac{1}{2}, x;\quad x, -x, \frac{1}{2}.$
(tilted carbonate groups)			with $x = -0.039$
3 O_I	e	2	$\frac{1}{2}, x, -x;$ etc.
(basal plane carbonate group)			with $x = 0.365$
3 O_{II}	e	2	$\frac{1}{2}, x, -x;$ etc.
(in-plane oxygens of tilted carbonate groups)			with $x = 0.096$
9 O_{III}	f	1	$x, y, z;$ etc. (see Table 8a)
(out-of-plane oxygens of tilted carbonate groups)			with $x = -0.033$
			$y = 0.180$
			$z = 0.371$

Cf. explanation to Table 5a.

Table 11b. Atomic sites in huntite structure after GRAF and BRADLEY (1962).
Space group $R32 - D_3^7$ (No. 155; International Tables I)
Occupancy of hexagonal unit cell

	Wyckoff notation	Point symmetry	Coordinates
3 Ca	a	32	$0, 0, 0.$
3 C_I	b	32	$0, 0, \frac{1}{2}.$
(basal plane carbonate group)			
9 Mg	d	2	$x, 0, 0;\quad 0, x, 0;\quad -x, -x, 0.$
			with $x = 0.541$
9 C_{II}	e	2	$x, 0, \frac{1}{2};\quad 0, x, \frac{1}{2};\quad -x, -x, \frac{1}{2}.$
(tilted carbonate groups)			with $x = 0.461$
9 O_I	e	2	$x, 0, \frac{1}{2};$ etc.
(basal plane carbonate group)			with $x = -0.135$
9 O_{II}	e	2	$x, 0, \frac{1}{2};$ etc.
(in-plane oxygens of tilted carbonate groups)			with $x = -0.404$
18 O_{III}	f	1	$x, y, z;$ etc. (see Table 8b)
(out-of-plane oxygens of tilted carbonate groups)			with $x = 0.461$
			$y = 0.135$
			$z = 0.506$
			$+ (0, 0, 0;\quad \frac{1}{3}, \frac{2}{3}, \frac{2}{3};\quad \frac{2}{3}, \frac{1}{3}, \frac{1}{3})$

Cf. explanations to Tables 5b and 1b.

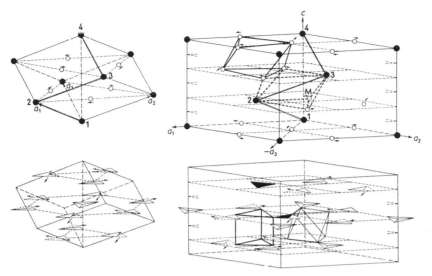

Fig. 20. The rhombohedral and hexagonal unit cells of huntite are shown as separate duplicates for the cations and the CO_3 groups. The orientation of the two cell types relative to each other is shown by the position of the common line 1-2-3-4. Small open circles are Mg; larger filled circles, Ca; solid-line triangles, carbonate groups lying in basal planes. The partially dotted triangles represent carbonate groups tilted out of basal planes about the indicated angular bisectrices as axes, with the dotted portion in each case lying below the basal plane. Small arrows indicate the hexagonal axial directions along which Mg are shifted from ideal cation positions. Arrows similarly show directions of shifts of the C of tilted carbonate groups. The C shifts have been ignored in drafting the rhombohedral cell. – Heavy solid lines connect the six Mg that surround a basal-plane carbonate group, colored black for reference. Dashed double lines similarly indicate the four Mg and two Ca that surround a tilted carbonate group, half-black for reference. Another set of heavy solid lines connects the six O forming a nearly right trigonal prism about a Ca atom ('2'). Thin solid lines outline the six O, forming an octahedron about a Mg atom ('M'); the two shared edges are shown as heavier lines (from GRAF and BRADLEY, 1962)

The hexagonal aspect of the structure, which is illustrated in the right part of Fig. 20, exhibits an alternation of cation levels as shown in Fig. 19a and CO_3 levels. Although the CO_3 groups are confined to basal levels as in the less complicated rhombohedral carbonates discussed above, they are no longer strictly coplanar as in these. Only the CO_3 group composed of C_I and O_I (as defined in Tables 11a and b), which is bonded solely by magnesium, is oriented parallel to the basal plane. Of the other CO_3 groups, bonded both by magnesium and calcium, only the carbon C_{II} and one oxygen O_{II}, both located on binary axes, are situated in the same basal plane as C_I and O_I. By their other two oxygens O_{III},

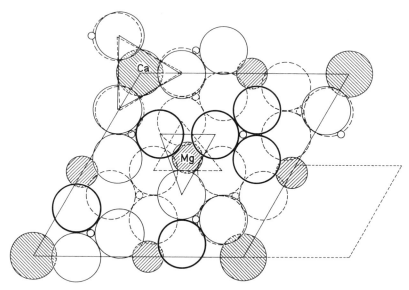

Fig. 21. Fundamental layer of huntite structure projected in the direction of the hexagonal c axis. The cations (shaded circles) have the coordinate $z = 0$. The carbons and the heavy oxygen circles of the CO_3 layer (full lines) on top of the cations have $z = \frac{2}{3} - \frac{1}{2} = \frac{1}{6}$. The in-plane CO_3 group has three heavy oxygens, O_I. Of the tilted CO_3 groups, the in-plane O_{II} is drawn heavy; the two out-of-plane oxygens, O_{III}, have normal lines. The O_{III} near Ca are tilted up by rotation about binary axes, which are defined by carbon and the centers of the heavy oxygen circles. The different types of oxygen are not distinguished in the lower CO_3 layer (broken lines; $z = \frac{1}{3} - \frac{1}{2} = -\frac{1}{6}$) but may be identified by considering the binary axis coinciding with the short diagonal of the hexagonal cell outline ($a_{hex} = 9.505$ Å). A calcite-type sub-cell is indicated by broken lines to the right. – The twisted oxygen prism around Ca is shown by its basal triangles. The triangles shown of the asymmetric octahedron around Mg are slightly tilted out of the basal plane as their acute corner, defined by out-of-plane O_{III}, is closer to the cation level than the other two corners, which are defined by (heavy) in-plane oxygens, O_{II} and O_I

these CO_3 groups are tilted out of the basal plane in a symmetrical fashion and in such a sense that the O_{III} coordinated to magnesium is closer to the cation level than the O_{III} coordinated to calcium. This tilt of the CO_3 groups and the coordination features of the cations are shown in Figs. 20 and 21. Calcium has six oxygen neighbors situated at the corners of a slightly twisted trigonal prism. The coordination octahedron around magnesium is highly asymmetric, its only symmetry operator being one binary axis. There are thus three types of Mg–O bonds and seven different O–O edges. The bond lengths of the huntite structure are listed in Table 12. Except for the extremely short O_I–O_{II} length of 2.56 Å, which is

Table 12. Interatomic distances (in Å) in huntite structure
after GRAF and BRADLEY (1962)

In basal plane:	CO_3 group	$C_{I,II}-O_{I,II,III}$	1.28
	Ca–O prism	$O_{III}-O_{III}$	3.32
	Mg–O octahedron	O_I-O_{II}	2.56 (the shared edge)
		O_I-O_{III}	3.10
		$O_{II}-O_{III}$	2.89
Out of basal plane:	Ca–O prism	$Ca-O_{III}$	2.35
		$O_{III}-O_{III}$	2.70
	Mg–O octahedron	$Mg-O_I$	2.10
		$Mg-O_{II}$	2.09
		$Mg-O_{III}$	2.10
		O_I-O_{II}	3.09
		O_I-O_{III}	3.06
		$O_{II}-O_{II}$	2.86
		$O_{III}-O_{III}$	3.30
Average O–O length of Mg–O octahedron			2.96

an edge shared by adjacent Mg–O octahedra, they are consistent with the interatomic distances known from the less complicated carbonate structures.

The averages of the hexagonal lattice constants of calcite and magnesite are slightly greater than the dimensions observed for the hexagonal cell of dolomite (see Table 14). This mineral is thus denser than a mixture composed of equal moles of calcite and magnesite. In contrast, lattice constants obtained for the huntite composition by linear interpolation between calcite and magnesite are significantly less than the dimensions actually observed for huntite. Accordingly, as pointed out by GRAF and BRADLEY, huntite has to be regarded as a low-density phase which would certainly not be stabilized by pressure. This is in harmony with the environmental conditions of the known huntite occurrences and would also explain why the mineral has never been obtained in hydrothermal experiments on compositions from which it might be expected.

3. Disordered Calcite-Type Structures

a) Magnesian Calcites

It has been known for a long time, at least since FORCHHAMMER (1852), that the hard parts secreted by certain organisms and consisting mainly of calcium carbonate may contain considerable amounts of magnesium carbonate in addition. Before the application of X-ray diffraction to the problem there was doubt as to the phase in which the magnesium might occur, although CLARKE and WHEELER (1924) had suggested that the magnesium carbonate present in skeletal parts must be somehow associated with calcite and that aragonitic organisms contained little or no magnesium. The authors arrived at this conclusion by correlating their chemical analyses of more than 300 specimens with the findings of MEIGEN (1903) who had differentiated calcite and aragonite in organic parts by means of his staining test using cobalt nitrate solution. The present state of knowledge concerning magnesium contents and mineralogical composition in various groups of organisms is summarized in Table 13. More data have been reviewed by DODD (1967).

X-ray diffraction studies of skeletal materials have demonstrated conclusively that the magnesium carbonate occurs in solid solution with calcite (CHAVE, 1954; GOLDSMITH, GRAF and JOENSUU, 1955). The latter group of authors have pointed out that these magnesian calcites are not restricted to organisms but may also be of purely inorganic origin when they occur as rock-forming constituents of metamorphic rocks, such as dolomitic marbles, and of calcareous tufa in the sedimentary cycle. The

Table 13. Magnesium carbonate in skeletal materials, compiled from CHAVE (1954) by GRAF (1960)

Classification	Number of samples	Percent aragonite	Weight percent $MgCO_3$
Foraminifera	23	0[a]	< 4 −15.9
Sponges	3	0	5.5 −14.1
Madreporian corals	10	100	0.12− 0.76
Alcyonarian corals	14	0	6.05−13.87
Echinoids	25	0	4.5 −15.9
Echinoid spines	12	0	< 4 −10.2
Asteroids	9	0	8.6 −16.17
Ophiuroids	6	0	9.23−16.5
Crinoids	22	0	7.28−15.9
Annelid worms	12	0− 99	6.40−16.5
Pelecypods	11	0−100	0.09− 2.80
Gastropods	7	5−100	0.08− 2.40
Cephalopods	3	0−100	0.05− 7.00
Decapod crustaceans	6	0	5.2 −11.70
Ostracode crustaceans	6	0	< 4 −10.2
Barnacles	9	0	1.35− 4.60
Calcareous algae	15	0[b]	7.7 −28.75

[a] In two families of foraminifera aragonite tests have been found by BLACKMON and TODD (1959).

[b] Calcitic forms only (e.g. Lithothamnium, Lithophyllum, Corallina). Aragonitic forms (e.g. Halimeda, Acetabularia), all of which are free from additional calcite, contain about 1% $MgCO_3$. See SCHOPF and MANHEIM (1967) for bryozoa. They may contain calcite and/or aragonite.

magnesian calcites contained in recent and sub-recent marine sediments
have been introduced predominantly as organic detritus, and they tend
to disappear with increasing geologic age. As a rule, they are no longer
present in solidified limestones of about Tertiary age and older, not even
in cases where these are composed essentially of organic remains with
inferred high primary magnesium contents (e.g. echinoderms, calcareous
algae).

Magnesian calcites are distinguished from pure calcite by the position of their X-ray reflections. These are shifted, with reference to their
positions in pure calcite, to higher diffraction angles, which are equivalent to smaller lattice spacings. This is what one would expect when the
cation sites of the calcite structure are partially occupied by magnesium
instead of calcium, given the fact that the lattice dimensions of dolomite
and magnesite are smaller than those of calcite. Correlations of chemically determined magnesium contents and measured X-ray spacings
have shown that the lattice dimensions of magnesian calcites vary with
the molecular percentage of magnesium in an essentially linear fashion
and thus follow VEGARD's rule. The relation between composition and
the hexagonal a and c dimensions is illustrated in Figs. 22 and 23, which
are from GOLDSMITH, GRAF and HEARD (1961). The discrepancies shown
for two different series of data are perhaps indicative of the experimental
difficulties inherent in the preparation of the samples measured, which
the authors carried out by quenching suitable compositions at high
temperatures and pressures [see GOLDSMITH and HEARD (1961) to be
discussed on p. 151]. In determining the composition of magnesian calcites from X-ray diffraction, approximate magnesium concentrations
may be deduced by linear interpolation between calcite and dolomite.
For this purpose, not only the lattice constants a_{hex} and c_{hex} but any
suitable lattice spacing may be used. Maximum sensitivity is obtained
from back reflections (see e.g. GRAF, 1961), but with small amounts of
magnesian calcite, the strongest reflection (10.4) often has to be used as
the last resort.

For higher accuracy, the results may be refined by considering
the data of Figs. 22 and 23. However, for such demands, direct chemical
determination is to be preferred, also in view of the possible presence in
the solid solution of cations other than Ca and Mg, such as Fe and Mn.
Furthermore, it is possible that the lattice dimensions of magnesian
calcites are not determined exclusively by the amount of magnesium
present. They may depend in addition to some slight extent upon the
mode of preparation or formation. Different types of disorder and CO_3
tilt (see p. 41) suggest themselves as possible complicating factors.

In summary, magnesian calcites may be regarded as minerals intermediate between calcite and dolomite not only with respect to their
chemical composition, but also their lattice dimensions. However, their

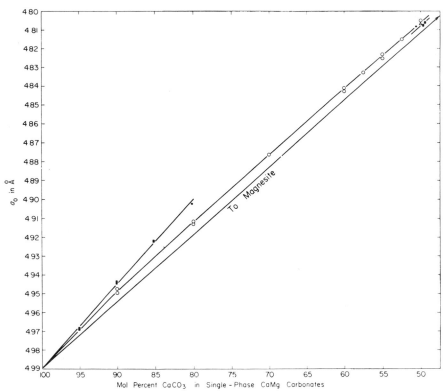

Fig. 22. The change of the hexagonal lattice constant $a_0 (= a_{hex})$ with composition in magnesian calcites (from GOLDSMITH, GRAF and HEARD, 1961). – Except for the value for (cation-ordered) dolomite, which is shown by a triangle, the data were determined on artificial magnesian calcites obtained by quenching appropriate compositions at high temperatures. The open circles are for materials prepared by GOLDSMITH and HEARD (1961). They are preferred by the authors over previous measurements (filled rectangles) by GOLDSMITH and GRAF (1958a), although no explanation for the discrepancy has been given. It is seen that a straight-line relationship is closely followed
(Reprinted with permission from *The American Mineralogist*)

kinship with dolomite does not include cation order, since superstructure reflections are absent in all organic carbonates. The highest magnesium contents occur in calcareous algae, and the value of about 30 mol% (Table 13) dates back to CLARKE and WHEELER. In contrast, the highest values so far confirmed by X-ray diffraction are about 20 mol% $MgCO_3$ in the magnesian calcites of calcareous algae. In particular, this is true in the case of the alga Goniolithon studied by SCHMALZ (1965). This author determined by X-ray diffraction that 18 mol% $MgCO_3$ are contained in the magnesian calcite, whereas chemical analyses yield values from 24 to

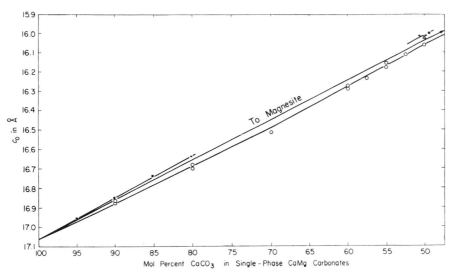

Fig. 23. The change of the hexagonal lattice constant $c_0 (= c_{hex})$ with composition in magnesian calcites (from GOLDSMITH, GRAF and HEARD, 1961). – The symbols have the same meaning as in Fig. 22
(Reprinted with permission from *The American Mineralogist*)

29 mol%. In the diffraction patterns SCHMALZ found the three strongest X-ray reflections of brucite, $Mg(OH)_2$, in addition to those of magnesian calcite. At least in the case of Goniolithon, the magnesium in excess of the amount contained in magnesian calcite is explained by the presence of brucite.

In view of the low relative intensities of the order reflections even of ideal dolomite, it is conceivable that a possible concentration of magnesium in every second cation layer does not show up in X-ray patterns of skeletal carbonates. However, there is no compositional gap between these and the phases with more than 40 mol% $MgCO_3$ in which superstructure reflections do occur (see below). Magnesian calcites between 20 and 40 mol% have been reported from the Tertiary of the Swiss Jura by KÜBLER (1958, 1962) and have been synthesized by various authors (see pp. 165—166). In all these cases order reflections did not appear. It is therefore generally accepted that magnesian calcites are essentially disordered, i.e. that their magnesium is randomly distributed in the cation sites of a calcite-type structure.

b) Calcium-Rich (Calcian) Dolomites

Minerals containing magnesium above 40 mol% occur with increasing frequency as they approach 50%, the ideal composition of dolomite.

In a detailed X-ray study on both powders and single crystals, GOLD-
SMITH and GRAF (1958) observed the presence of order reflections,
though of varying intensity, in the majority of the natural samples they
examined in this compositional range. For the determination of the
deviations from the ideal dolomite composition they assumed a straight-
line relationship between dolomite and calcite in the plot of lattice spacing versus molar composition. This procedure has been followed by later
investigators (e.g. FÜCHTBAUER and GOLDSCHMIDT, 1965). It may be
seen from Figs. 22 and 23 that such straight lines through the points
representing dolomite must be virtually parallel to the line for the disordered phases and to the join calcite-magnesite in the range considered.
These different lines are separated according to the differently defined
lattice dimensions for the 50:50 composition in Table 14. GOLDSMITH

Table 14. Hexagonal lattice constants (in Å) for the composition
$CaCO_3 : MgCO_3 = 50 : 50$ (from GOLDSMITH, GRAF and HEARD, 1961)

	a	c
Cation-ordered dolomite	4.8079	16.010
Disordered by heating to 1155° C	4.8050	16.045
Synthetic, disordered	4.8050	16.061
Arithmetic mean for magnesite and calcite	4.8114$_5$	16.039$_5$

and GRAF (1958) observed maximum intensities of the order reflections
for those dolomites which deviate least from the ideal composition. No
sample showed a significant excess of magnesium. However, a substantial number of the samples turned out to be high in calcium with a
frequency maximum at about 55 to 56 mol% $CaCO_3$. An indication of
such a maximum is apparent also in the data of FÜCHTBAUER and
GOLDSCHMIDT (1965), but the general impression is that of a continuous
series of magnesium deficient dolomites ranging from the ideal composition to about 40 mol% of magnesium. Both groups of authors established the general tendency that the superstructure reflections decrease
more or less drastically with increasing excess of calcium. GOLDSMITH
and GRAF (1958) calculated the relative intensities of superstructure reflections for a composition of 55 mol% Ca, assuming accommodation of
the excess calcium in the magnesium layer of the dolomite structure. The
calculated decrease in intensity was smaller than the observed attenuation of the order reflections. The disorder due to an excess of calcium is
thus incompletely described by a simple substitution of a small portion
of the magnesium by calcium. In addition to the attenuation of the order

reflections, GOLDSMITH and GRAF (1958) observed that the basal spacings of the type (00.*l*) also tend to become weaker and broader in calcium-rich dolomites. The same is true for reflections *hk.l* with relatively high *l*. In some cases 00.*l* appeared to be doubled whereas reflections with *l* = 0 remained sharp. To account for these observations, the authors proposed to explain the excess in calcium by a somewhat irregular interstratification of magnesite-like and calcite-like layers, with an excess of the latter. This picture was obviously suggested by the irregular mixed-layer structures occuring in clays. GRAF, BLYTH and STEMMLER (1967) carried out detailed calculations for simple as well as more complicated interstratification models and arrived indeed at broadened and also bimodal 00.*l* reflections. The basis for the assumption of irregularly interstratified models has been the relative sharpness of the *hk.*0 reflections. However, when X-ray diffractometer traces of different dolomites are recorded under the same conditions it turns out that all reflections are weaker in the magnesium-deficient dolomites than they are in the stoichiometric ones, including the lines indexed *hk.*0 and others, which should not be affected by irregular interstratification (LIPPMANN, unpublished). The fact remains, however, that basal and order reflections are weakened to the greatest extent. In the disordered models considered so far, the CO_3 groups have been assumed to be coplanar as in dolomite. However, when the same cation layer contains both cations, as has also been considered for certain models by GRAF, BLYTH and STEMMLER, one will expect that at least part of the CO_3 groups are tilted as in huntite. The tilting will enhance the variability of the c_{hex} spacing, which is more sensitive already to cation size than the a_{hex} dimension, as shown by the lattice parameters of the pure carbonates. This would help to explain the broad 00.*l* reflections by variable composition in the same crystallite, i.e. by crystal zoning. Since the CO_3 tilt affects both cation layers adjacent to a given CO_3 layer, it is unlikely that a calcium substitution in the magnesium layers is compatible with pure calcium layers. It is conceivable that a substitution in the magnesium layer entails a certain substitution also in the calcium layer. In this way, certain compositions might be favored as e.g. the 45:55 ratio. Intermediate ratios may be composed of discrete zones with favored ratios, among them the ideal 1:1 ratio, and this might be an alternative explanation for doubled 00.*l* reflections. A randomly distributed tilt of the CO_3 group will be reflected in anisotropic temperature factors. Therefore, crystal structure refinements, including the analysis of the apparent thermal motion of the oxygen, appear highly desirable for calcium-rich dolomites.

The study of GOLDSMITH and GRAF (1958) included 38 samples from dolomitic rocks of different geologic ages. All of the well-ordered dolomites were from evaporite environments whereas normal marine formations seemed to be characterized throughout by disordered dolomites showing excess of calcium. FÜCHTBAUER and GOLDSCHMIDT found this relationship to be true statistically, but since they studied a greater number of samples they also found a great many exceptions, i.e. ideally ordered dolomites occurring in non-evaporite environments. FÜCHT-BAUER and GOLDSCHMIDT have pointed out other relationships concerning the occurrence of ideally composed versus magnesium-deficient dolomites. The latter appear to favor rocks which contain calcite in addition and which form layers interstratified with, or patches enclosed in, limestone, i.e. such sedimentary structures as are usually regarded as indicative of primary or early diagenetic dolomite formation. In contrast, ordered dolomites appear to form essentially pure bodies which are of greater dimensions and more massive in character, and which are usually referred to as being of late diagenetic origin. Exceptions can be found also to these rules. Dolomite bands of several centimeters thickness which are interstratified with the limestone of the Upper Muschelkalk of Southern Germany (LIPPMANN and SCHLENKER, 1970) contain essentially ideal dolomite. Even so, the generalizations of FÜCHTBAUER and GOLDSCHMIDT are in line with the fact that most dolomites found in recent sediments (to be discussed on p. 149), which have thus formed in early diagenesis, are magnesium-deficient and rarely exhibit superstructure reflections.

4. Iron-Bearing Rhombohedral Carbonates

When we consider the overall abundance of carbonate minerals in sedimentary rocks, calcite and dolomite are followed by siderite and iron-bearing dolomites (ankerites) in order of decreasing importance. These minerals are in general not deposited in primary sediments as they require reducing conditions for their formation, on account of the bivalent iron they contain. It is generally accepted that they form during diagenesis under reducing conditions and so may be used as indicators of such conditions. Therefore, their importance in sedimentary petrology becomes enhanced in comparison to their abundance.

a) Siderites

If the siderites occurring in sedimentary rocks were essentially pure $FeCO_3$, it would be sufficient to know that $FeCO_3$ is isostructural with calcite and to characterize it by its lattice constants given in Table 3. However, chemical analyses available for siderites from sedimentary formations (e.g. BORNMÜLLER, 1923; SMYTHE and DUNHAM, 1947; TAYLOR,

1950) show considerable amounts of calcium and magnesium. Virtually all of the available chemical data for such minerals are not supplemented by X-ray data. An X-ray study of materials similar in occurrence to those analysed by BORNMÜLLER, i.e. of septarian nodules from the Lower Cretaceous shales of the surroundings of Hannover, Germany, has shown that the carbonate part of the siderite nodules is monomineralic in most cases (LIPPMANN, unpublished). There is thus reason to conclude that the calcium and the magnesium, reported for such ironstones in magnitudes of about 10 mol% each and more, are incorporated in the siderite in solid solution. Instead of the (10.4) spacing for pure siderite at 2.791 Å, most of the samples studied showed values around 2.81 Å. In view of the small deviation in one direction of the corresponding magnesite spacing (2.74 Å) and the greater deviation of calcite (3.03 Å) the other way, it is to be expected that the spacings of solid solutions with about the same amounts of magnesium and calcium are shifted in the direction of calcite. Assuming that according to VEGARD's rule, linear interpolation between the pure end-members is permissible, we may write the following equation for the (10.4) spacing of siderite mixed-crystals

$$d_{10.4} = 2.791 - 0.050\,x + 0.245\,y$$

where x is the mol fraction of $MgCO_3$ and y that of $CaCO_3$. Since the equation contains two unknowns it is not possible to determine the composition from the spacing alone. However, estimates may be obtained by recourse to other data. LIPPMANN (1955) proposed considering the refractive index n_ω, and MORELLI (1967) has used calculated X-ray diffraction intensities as additional information. From the combination of X-ray spacings and intensities, the latter author has obtained estimates which compare favorably with the chemical analyses of four natural siderite mixed-crystals of various origin. More information, derived from combined X-ray and chemical studies, is needed for a more complete understanding of the variability of the siderites occurring in sedimentary formations.

b) Iron-Bearing Dolomites or Ankerites

$CaFe(CO_3)_2$, the hypothetical iron analogue of dolomite, has never been found in nature as a pure compound, nor has it formed in the experiments of GOLDSMITH, GRAF, WITTERS and NORTHROP (1962) on the system $CaCO_3$–$MgCO_3$–$FeCO_3$. However, dolomites in which part of the magnesium is substituted by bivalent iron, and which are usually referred to as ankerites, occur with some frequency as rock-forming minerals. WYCKOFF and MERWIN (1924), who carried out the first X-ray study of such a mineral intermediate between dolomite and $CaFe(CO_3)_2$,

4*

obtained reflections analogous with pure dolomite and confirmed that ankerites are isostructural with dolomite, as had been concluded before from chemical analyses.

The maximum possible substitution of magnesium by iron in dolomite-type minerals pertaining to the sedimentary cycle is not known for certain. Even in cases where ankerites are associated with sedimentary rocks, as e.g. those analysed chemically by SMYTHE and DUNHAM (1947), it is not always clear whether a certain composition has formed by diagenetic processes, and thus belongs to the sedimentary cycle, or is a product of later epigenetic deposition. In general, ankerite compositions show a broad frequency maximum about half-way between $CaMg(CO_3)_2$ and $CaFe(CO_3)_2$, and one may get the impression that sedimentary ankerites are lower in iron than the average, from such isolated examples as the mineral with 16.6 mol% $FeCO_3$, described by HOWIE and BROADHURST (1958) from the British coal measures.

GOLDSMITH et al. (1962) have carried out an X-ray study by the powder method of 20 ankerites of various origin. The correlation with chemical composition was complicated by the presence of manganese of the order of about 1 mol% in most samples. The authors deduced the following linear regression formulae interrelating chemical composition and hexagonal lattice constants:

$$a_{hex} = 4.92954\,N_{CaCO_3} + 4.6929\,N_{MgCO_3} + 4.73269\,N_{FeCO_3}$$
$$+ 4.71879\,N_{MnCO_3}$$

and

$$c_{hex} = 16.5368\,N_{CaCO_3} + 15.5004\,N_{MgCO_3} + 15.8589\,N_{FeCO_3}$$
$$+ 16.0111\,N_{MnCO_3}$$

where N_{XCO_3} is the mol fraction of XCO_3. The formulae are valid only for ankerite mixed-crystals and will certainly lead to erroneous results outside this compositional range, since the coefficients differ considerably from the lattice constants of the pure rhombohedral carbonates (cf. Table 3). It is likely that these deviations reflect something of the special spatial requirements due to the cation order in ankerite structures.

The intensity changes with increasing iron content of the dolomite-type order reflections have been described by GOLDSMITH et al. (1962) for artificial ankerites synthesized at 700° C and 800° C. The order reflections (10.1), (01.5) and (02.1) become weaker with increasing iron substitution. The first of these reflections to disappear is (10.1), above about 10 mol% $FeCO_3$ [= 20 mol% $CaFe(CO_3)_2$], whereas (01.5) and (02.1) are no longer observed above 25 mol% $FeCO_3$ [= 50 mol% $CaFe(CO_3)_2$]. The gradual attenuation of these order reflections shows that the iron enters the Mg levels of the dolomite structure. Their scattering power approaches that of the calcium levels with increasing iron incorporation. Since it is the difference in scattering power which determines the intensities of the order reflections, the gradual decrease of the latter can be explained. That they disappear at different iron contents is due to different oxygen contributions. GRAF (1961) has substantiated this general behavior of the superstructure reflections by intensity calculations for ordered rhombohedral compositions between $CaMg(CO_3)_2$ and $CaFe(CO_3)_2$.

When interpreting the X-ray pattern of some unknown material, the decrease of the superstructure reflections under the influence of iron substitution may be confused with a similar decrease due to cation disorder in iron-free dolomite (described under calcium-rich dolomites, p. 48). In such a case the (00.3) reflection is of use. Owing to the sign of the oxygen contribution, (00.3) is too weak to be detected in powder diffraction from ideal $CaMg(CO_3)_2$, and it is self-evident that it will be further weakened by cation-disorder. However, it is enhanced by the incorporation of iron in the Mg level of ordered dolomite structures according to the calculations of GRAF (1961). GOLDSMITH et al. (1962) have observed it in their artificial ankerites for iron contents above about 15 mol% of $FeCO_3$. Therefore, the consideration of the (00.3) reflection and the fact that the other superstructure reflections do not disappear simultaneously with increasing iron substitution may help to distinguish X-ray patterns of ankerites and iron-free disordered dolomites. Additional complications arise with dolomite-type minerals which contain iron and, at the same time, show an excess of calcium over the formula $(Mg, Fe)Ca(CO_3)_2$. Some of the minerals studied by GOLDSMITH et al. (1962) are just of this type. Their existence shows that X-ray diffraction can afford nothing more than a preliminary determination of an ankerite and that complete characterization is possible only in conjunction with quantitative chemical analyses. More studies along these lines are needed before the chemical and structural variability of ankerites in sedimentary and other environments can be fully appreciated. Moreover, independent crystal structure refinements should be carried out for a number of characteristic representatives.

II. Aragonite

Although aragonite, the second polymorph of $CaCO_3$, is not as important in overall abundance as calcite, it plays an important rôle in the formation of carbonate rocks, in that it is a main constituent of many recent marine carbonate sediments. Into these, it is introduced both as a product of direct precipitation from sea-water and as the detritus of the hard parts of certain organisms (see Table 13). Aragonite is rarely found in sediments of Tertiary age or older. Even so, from the abundance of aragonite in recent calcareous sediments and also from microscopic criteria (SORBY, 1879; BØGGILD, 1930; HOROWITZ and POTTER, 1971) it must be concluded that most ancient marine carbonate rocks, which to-day consist of calcite and/or dolomite, contained substantial amounts of aragonite when they were deposited. In the evolution of marine carbonate rocks, the mineral appears to characterize the interval extending from deposition to some time prior to the final lithification. So we may hope to understand at least something of the depositional and diagenetic processes involved, if we are able to explain the conditions of formation and transformation of aragonite. In trying to do so later in part C, we shall have to rely to a considerable extent on the structural information to be discussed in the following paragraphs.

The structural scheme of aragonite was determined from X-ray diffraction by W. L. BRAGG (1924). Quite recently, this analysis has been confirmed by DAL NEGRO and UNGARETTI (1971), who applied the modern techniques of crystal structure refinement.

On p. 6 ff.; it has been shown how the structure of calcite may be derived from the NaCl arrangement. In an analogous fashion, according to EWALD and HERMANN (1931), aragonite is related to another type of a

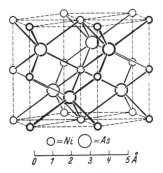

$O = Ni$ $\bigcirc = As$

$\underset{0\quad 1\quad 2\quad 3\quad 4\quad 5\,\text{Å}}{\vdash\!\!\!\!\dashv}$

Fig. 24. The crystal structure of niccolite, NiAs (from CORRENS). The lattice constants are: $a = 3.58$ Å; $c = 5.11$ Å. The components occupy the following equipoints of the space group $P6_3/mmc - D_{6h}^4$ (No. 194, International Tables I):

	Wyckoff notation	point symmetry	Coordinates
2 Ni	a	$\bar{3}m$	0, 0, 0; 0, 0, $\frac{1}{2}$.
2 As	d	$\bar{6}m2$	$\frac{1}{3}$, $\frac{2}{3}$, $\frac{3}{4}$; $\frac{2}{3}$, $\frac{1}{3}$, $\frac{1}{4}$.

To obtain the structural scheme of aragonite, As is thought to be replaced by Ca, and Ni by CO_3

simple binary compound, viz. niccolite, NiAs, whose structure is illustrated in Fig. 24. Ni is surrounded octahedrally by six As, but in contrast to the mutual octahedral coordination of both cations and anions in NaCl, As in niccolite is surrounded by six Ni which are situated at the corners of a trigonal prism. By assuming a replacement of As by Ca and of Ni by CO_3 groups, we obtain the general pattern of aragonite. The CO_3 groups are coplanar and their planes are perpendicular to the c axis of the original niccolite structure, which coincides with the conventional c axis of the resulting aragonite structure. According to the niccolite pattern, Ca and CO_3 should be arranged in layers perpendicular to the c axis. In actual aragonite the CO_3 layers are not flat, because CO_3 groups in adjacent rows parallel to the a axis have different z coordinates (Fig. 25), so that the CO_3 layers appear corrugated. Hence, the coordination prism around Ca, with the centers of CO_3 groups at its corners, which corresponds to the trigonal $AsNi_6$ prism in niccolite, is no longer a right prism. It is truncated in an oblique fashion by its basal planes which, nevertheless, remain mutually parallel. This deformation is certainly more radical than the homogeneous compression which leads from the NaCl pattern to calcite. Therefore, we may not expect to obtain more structural details from the comparison to niccolite. Not even the trigonal rhythm of the hexagonal screw-axis 6_3 of niccolite is retained to control the orientation of the CO_3 groups in aragonite. Only the tendency of the mineral to form pseudohexagonal twins, which BRAGG (1924) was able to interprete on the basis of his crystal structure (see Fig. 28) is reminiscent of the structural kinship with niccolite as well as of the pseudohexagonal character of the whole structure.

In comparison to the hexagonal space group $P6_3/mmc - D_{6h}^4$ of niccolite, the symmetry of aragonite in the orthorhombic space group $Pmcn - D_{2h}^{16}$ appears considerably lowered. The atomic coordinates of this space group, according to International Tables I and permuted from $Pnma$ to $Pmcn$ to conform with the conventional choice of axes a, b, and c are listed in Table 15a, along with the positional parameters of BRAGG's structure model. The latter has been found to be essentially correct by DAL NEGRO and UNGARETTI (1971) as shown in Table 15b where the refined parameters of these authors are compared with those for the BRAGG model. Therefore, we may continue using the BRAGG model in visualizing the general features of the structure, but we shall have to refer to the refined structure when more detailed information is required. In particular, for the interatomic distances, (Tables 16 and 17) we shall rely entirely upon the data of DAL NEGRO and UNGARETTI.

In the space group $Pmcn$, the orientation of the CO_3 groups is governed mainly by the location of their carbon centers on mirror planes parallel to b and c. Besides, by the action of binary screw axes parallel to

Table 15a. Atomic sites in aragonite structure after W. L. BRAGG (1924).
Space group $Pmcn(Pnma) - D_{2h}^{16}$ (No. 62; International Tables I)

	Wyckoff notation	Point symmetry	Coordinates
4 Ca	c	m	$\frac{1}{4}, \quad y, \quad z; \quad \frac{1}{4}, \frac{1}{2} - y, \frac{1}{2} + z;$ $\frac{3}{4}, -y, -z; \quad \frac{3}{4}, \frac{1}{2} + y, \frac{1}{2} - z.$ $y = 0.417 \cong \frac{5}{12}; \quad z = 0.250 = \frac{1}{4}$
4 C	c	m	$\frac{1}{4}, y, z; \quad$ etc. $y = 0.750 = \frac{3}{4}; \quad z = 0.083 = \frac{1}{12}$
4 O$_I$	c	m	$\frac{1}{4}, y, z; \quad$ etc. $y = 0.917; \quad z = 0.083 = \frac{1}{12}$
8 O$_{II}$	d	1	$\pm x, \pm y, \pm z; \quad \frac{1}{2} \mp x, \frac{1}{2} \mp y, \frac{1}{2} \pm z;$ $\frac{1}{2} \pm x, \mp y, \mp z; \quad \mp x, \frac{1}{2} \pm y, \frac{1}{2} \mp z;$ $x = 0.480; \quad y = 0.670; \quad z = 0.083 = \frac{1}{12}$

The numerical values for the coordinates are those of BRAGG's model. In the original paper (1924) and also in BRAGG and CLARINGBULL (1965) as well as in Figs. 25 and 27, the origin has been shifted by $\frac{1}{4}$ along the c axis, i.e. $\frac{3}{4}$ or $-\frac{1}{4}$ have to be added to all z coordinates of this table in order to arrive at BRAGG's setting. EWALD and HERMANN (1931) have adopted a setting which, in addition to a $\frac{c}{4}$ shift, involves also a shift of the origin by $\frac{a}{4}$ with respect to the usage of International Tables I.

Table 15b. Numerical values for atomic coordinates (positional parameters) in aragonite structure. Comparison of refined structure (DAL NEGRO and UNGARETTI)[a] and BRAGG's model[b]

		x	y	z
Ca	$\{$	0.2500	0.4151(1)	0.2403(1)
		$(\frac{1}{4})$	(0.417)	$(\frac{1}{4})$
C	$\{$	0.2500	0.7627(5)	0.0850(6)
		$(\frac{1}{4})$	$(\frac{3}{4})$	$(\frac{1}{12} = 0.0833)$
O$_I$	$\{$	0.2500	0.9231(3)	0.0952(5)
		$(\frac{1}{4})$	(0.917)	$(\frac{1}{12} = 0.0833)$
O$_{II}$	$\{$	0.4729(4)	0.6801(3)	0.0870(4)
		(0.480)	(0.670)	$(\frac{1}{12} = 0.0833)$

x, y, and z are given as fractions of the cell lengths $a, b,$ and c, respectively.

[a] Values followed by standard deviations in parentheses.

[b] Values in parentheses.

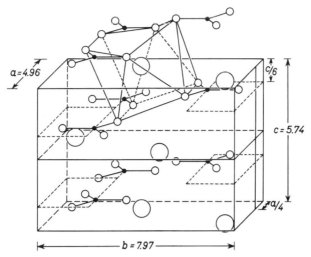

Fig. 25. Unit cell of aragonite, containing $4\,CaCO_3$. Additional CO_3 groups are shown outside the unit cell in order to obtain the complete surrounding of one of the calciums (large circles), for which the coordination polyhedron has been drawn. The latter is shown separately in Fig. 26. The origin and the coordinates used to prepare this drawing are those of BRAGG's (1924) original description. The origin according to the conventions of International Tables I, followed in Tables 15a and 15b, would be at $z = \frac{3}{4}(=-\frac{1}{4})$ of the c length here shown. The refined coordinates of DAL NEGRO and UNGARETTI (1971) would yield slightly deviating atomic positions. However, at the present scale, the difference would be barely perceptible

all three crystallographic axes a, b, and c; the orientations of the six CO_3 groups surrounding one particular Ca are distributed in such a way that the calcium has nine immediate oxygen neighbors. These are situated at the corners of a rather asymmetric polyhedron (shown in Figs. 25, 26 and 27) whose only symmetry element is the mirror plane parallel to b and c. It must be noted that this CaO_9 polyhedron occurs in two different orientations with respect to the c axis. Like-oriented polyhedra, linked by common CO_3 groups, are arranged in rows parallel to the a axis. These rows are connected with each other via common CO_3 groups and by shared polyhedral edges. In a given row, the polyhedra show an orientation which is inverted with respect to the polyhedra in immediately adjacent rows. For example, the polyhedron shown in Figs. 25 and 26 is inverted with respect to that outlined in Fig. 27. Zig-zag rows with polyhedra of alternating orientation, connected via shared edges and common CO_3 groups, run along the b and c axes.

The surrounding of calcium by six CO_3 groups and, at the same time, by nine oxygens implies that three CO_3 are bonded to the same calcium by two oxygens each (Figs. 25, 26 and 27). This situation is different from the CO_3–Ca bonding in calcite where a given CO_3 group is bonded to a given calcium via one oxygen only (Fig. 4). In every coordination polyhedron of the aragonite structure, one of the CO_3 contributing two oxygens is aligned in a symmetric position with respect to the central calcium. The oxygens are defined by the Ca–O distance of 2.54 Å (see Fig. 26). The two other CO_3 groups extending two oxygens each to the central calcium deviate moderately from the symmetric orientation with respect to the latter, as indicated by the different Ca–O distances of 2.53 Å and 2.66 Å. If these two bond lengths become equal, or nearly so, by way of some misalignment occurring in a growing crystal, the CO_3 concerned may take the rôle of the symmetrically aligned CO_3 and, if the crystal continues growing accordingly, the twinning shown in Fig. 28 is the result. It is supposed that the interchange in rôles between the symmetric and asymmetric CO_3 takes place in the twin boundary as indicated in Fig. 28.

The remaining three oxygens of the CaO_9 polyhedron belong to three different CO_3 groups, which thus contribute one oxygen each. Of these, two have Ca–O distances of 2.45 Å. The shortest Ca–O length in the structure is 2.41 Å, and the pertaining CO_3 is aligned symmetrically with respect to the polyhedron (and the central Ca).

All of the Ca–O distances in aragonite (Table 16), including the shortest of 2.41 Å, are substantially greater than the Ca–O distance in calcite of 2.36 Å (Table 2). This relation seems to be at variance with the fact

Table 16. Cation-oxygen distances (in Å) in refined structure of aragonite
(DAL NEGRO and UNGARETTI, unpublished)

	$Ca–O_{11}$ $= Ca–O_{(11)}$	2.656(1)
	$Ca–O_{12}$ $=$	2.411(3)
	$Ca–O_{II}$ $= Ca–O_{II3}$	2.542(2)
	$Ca–O_{II1} = Ca–O_{II4}$	2.526(2)
	$Ca–O_{II2} = Ca–O_{II5}$	2.448(2)
	average Ca–O	2.528
	$C–O_I$ $=$	1.280(5)
	$C–O_{II}$ $= C–O_{II3}$	1.287(3)
Angle	$O_{11}–C–O_{II1} = 120°42'(11')$	
Angle	$O_{II}–C–O_{II3} = 118°28'(21')$	

Based on the parameters shown in Table 15b and the lattice constants as newly determined by the authors:

$$a = 4.9616 \pm 0.0002; \quad b = 7.9705 \pm 0.0006; \quad c = 5.7394 \pm 0.0004$$

The locations of the individual atoms are shown in Fig. 26. Standard deviations are in parentheses.

that aragonite is denser than calcite. Although the greater density of
aragonite is commonly ascribed to the coordination of calcium by nine
oxygens instead of six in calcite, such an explanation is hardly satisfactory, since, in the end, the mutual surrounding of Ca and CO_3 by six
neighbors is the same in both cases.

In this connection, it is interesting to consider the oxygen-oxygen
edges of the CaO_9 polyhedron. Their lengths are given in Table 17 for the
refined structure. It turns out that the distances between oxygens belonging to different CO_3 groups are smaller than the corresponding O–O
distances of 3.26 Å and 3.41 Å in calcite (Table 2). The only exceptions
are the edges O_{III}–O_{12} and O_{II4}–O_{12} with 3.60 Å (see top of Fig. 26a). One
could reasonably disregard these edges and take the view of a coordina-
tion polyhedron which is open on that side (see Fig. 25). The other O–O
distances between different CO_3 groups range from 3.23 Å to 2.75 Å. This
latter value is similar to two ionic radii of oxygen, and such a close
approach of two oxygens is not uncommon for an edge which is shared
by two adjacent coordination polyhedra (cf. huntite, Table 12). Even if
the two long edges are included, the average of 3.17 Å for the O–O
distances between different CO_3 groups is smaller than the shortest O–O
edge (3.26 Å) in the CaO_6 octahedron in calcite. From this, it may be
concluded that the greater density of aragonite relative to calcite is due
to more efficient packing of the CO_3 groups rather than to the difference
in Ca–O coordination. Evidently, the shorter Ca–O bond in calcite is
overcompensated by more closely packed CO_3 groups in aragonite.

Although in the discussion of the interatomic distances we have used
the numerical values for the refined structure, we have so far ignored the
fact that the latter deviates slightly from BRAGG's model in certain details. As we have seen above, the CO_3 layers in aragonite, which correspond
to the Ni planes of niccolite, are distinctly corrugated. By assigning
the coordinate $z = \frac{1}{4} = 0.25$ to one of the calcium atoms BRAGG obtained
flat calcium layers which, in their turn, correspond to the basal As planes
in niccolite. The refinement by DAL NEGRO and UNGARETTI has yielded
$z = 0.2403$ for the calcium in question. In this way, $\frac{1}{2} - z$ yields a value
different from z, and the calcium layers, too, become slightly corrugyted,
the difference in elevation between two Ca adjacent in the unit call
amounting to 0.111 Å. For similar reasons, the CO_3 strings along c are
slightly staggered in the b direction (see legend to Fig. 27).

We have also tacitly assumed that the CO_3 groups are planar as in
calcite. The C–O distances and the corresponding bond angles in
Table 16 show indeed that the CO_3 group in aragonite has very nearly
the same dimensions and shape as that in calcite. In particular, the
lengths C–O_I of 1.280(± 5) Å and C–O_{II} of 1.287(± 3) Å almost coincide
with the C–O distance of 1.281 Å in calcite, thus justifying the assump-

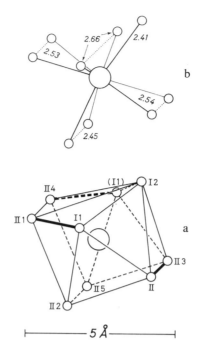

Fig. 26a and b. CaO$_9$ polyhedron in aragonite structure, showing coordination of calcium (large circle) by nine oxygens (small circles). The polyhedron is the same as that drawn in Fig. 25. a In the lower figure, which shows all of the oxygen-oxygen edges, individual oxygens are marked by the index numbers assigned in Tables 16 and 17, in order to identify the interatomic distance listed there. Pairs of oxygens belonging to one CO$_3$ group are joined by heavier lines. b In the upper figure, the Ca–O bonds are shown. They are marked by the Ca–O distances as determined by DAL NEGRO and UNGARETTI and listed in Table 16. Pairs of oxygens at equal distance from the central calcium are joined by dotted lines. – The short O–O edges, of 2.75 Å, shared with adjacent CaO$_9$ polyhedra, are O$_{II4}$–O$_{II1}$ and O$_{II5}$–O$_{II2}$. Other shared edges are O$_{II2}$–O$_{II}$ and O$_{II1}$–O$_{I2}$ as well as their symmetric counterparts O$_{II5}$–O$_{II3}$ and O$_{I2}$–O$_{(I1)}$

tion made on p. 17 that the CO$_3$ has approximately constant dimensions in different carbonates. However, in finer detail the z coordinates for the single components of a CO$_3$ group in the refined structure of aragonite are not only different from BRAGG's ideal value of $\frac{1}{12} = 0.0833$ but they also deviate from each other outside the limits of error (Table 15b). It is likely that the hereby indicated deviations from planarity in the CO$_3$ group help to facilitate the close approach of the oxygens belonging to different CO$_3$ groups, perhaps in combination also with the corrugation of the calcium layers and the b stagger of the CO$_3$ strings

Table 17. Oxygen-oxygen distances (in Å) in CaO_9 polyhedron (see Fig. 26) of refined aragonite structure (DAL NEGRO and UNGARETTI, unpublished)

Subhorizontal edges	above Ca	$O_{II4}-O_{II1}(= O_{II5}-O_{II2})$	2.750
		$O_{II1}-O_{I1} = O_{II4}-O_{(I1)}$	2.231
		(within CO_3 group)	
		$O_{I1}-O_{I2} = O_{(I1)}-O_{I2}$	2.975
		$O_{II1}-O_{I2} = O_{II4}\,O_{I2}$	3.603
	below Ca	$O_{II5}-O_{II2}(= O_{II4}-O_{II1})$	2.750
		$O_{II2}-O_{I1} = O_{II5}-O_{II3}$	3.051
		$O_{I1}-O_{II3}$ (within CO_3 group)	2.212
Subvertical edges		$O_{II1}-O_{II2} = O_{II4}\,O_{II5}$	3.078
		$O_{I1}-O_{II2} = O_{(I1)}-O_{II5}$	3.141
		$O_{I1}-O_{I1} = O_{(I1)}-O_{II3}$	3.068
		$O_{I2}-O_{I1} = O_{I2}-O_{II3}$	3.226

Cf. explanation to Table 16.
Standard deviations: 0.002–0.003 Å.

along c. These deviations from the BRAGG model may thus contribute to the more efficient packing of the CO_3 in aragonite and consequently to the greater density of the mineral in comparison to calcite. All these questions could perhaps be settled by calculations of the crystal ("lattice") energy for different aragonite models along the general lines of LENNARD-JONES and DENT (1927). Such work would certainly lead to a better understanding of certain points concerning the polymorphism calcite–aragonite (p. 98).

Thus far, we have considered the bonding in aragonite mainly from the point of view of the coordination of the calcium. However, to obtain a full grasp of the differences in arrangement with respect to calcite, we should see also the structure from the point of view of the CO_3 group as did BRAGG (1924) in his original description (see also BRAGG and CLARINGBULL, 1965). In both aragonite and calcite, a given CO_3 is surrounded by six calciums. In calcite, each oxygen is linked to two Ca, one of which belongs to the cation layer above and the other to the one below. The individual Ca touches only one oxygen of a given CO_3 group, and it occupies a position asymmetric with regard to the CO_3 group as a whole (Figs. 4 and 5). In aragonite, each oxygen is linked to three different calcium atoms (ions) (Fig. 28). This results from the alignment of three of the surrounding calcium atoms, belonging to one of the adjacent cation layers, in the direction of the interstices between the oxygens of the CO_3 group. These calcium atoms are thus bonded simultaneously by two oxygens of the CO_3 group. One of them occupies a strictly symmetric position with regard to the CO_3, which is characterized by two equal

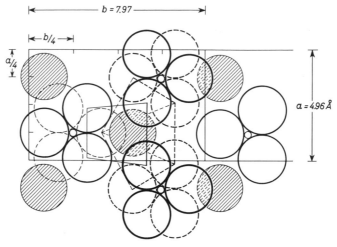

Fig. 27. Individual calcium carbonate layer of aragonite structure projected in the direction of the c axis on the (a, b) plane, after BRAGG (1924). The different z co-ordinates of the CO_3 groups, which lead to the corrugation of the CO_3 layers above (full circles) and below (broken circles) the Ca level (shaded circles), have the following values according to BRAGG if Ca has $z = \frac{1}{2}$:

$$\left.\begin{array}{ll} \dfrac{5}{6}c & \text{heavy full circles} \\[2mm] \dfrac{2}{3}c & \text{normal full circles} \end{array}\right\} \text{top layer,}$$

$$\left.\begin{array}{ll} \dfrac{c}{3} & \text{normal broken circles} \\[2mm] \dfrac{c}{6} & \text{light broken circles} \end{array}\right\} \text{lower layer.}$$

Of the oxygen polyhedron around one of the calciums, the edges subparallel to the projection plane are indicated. Note that the coordination polyhedron here is inverted with respect to the one in Fig. 25 where the central calcium belongs to the level above the one shown in the present figure. – In the refined structure, the carbons do not exactly project on top of each other as shown here, since the y coordinate of the carbon is not exactly $\frac{3}{4} = 0.75$ as assumed by BRAGG. Hence, the carbons are displaced relative to each other by about 0.2 Å along the b axis (see DE VILLIERS, Fig. 1, and DAL NEGRO and UNGARETTI, Fig. 1)

Ca–O distances of 2.54 Å each. The other two Ca have slightly asymmetric positions as indicated by two different Ca–O distances, 2.53 Å and 2.66 Å. In contrast, the three Ca, which belong to the cation layer adjacent to the CO_3 from the other side, are aligned in the projections of C–O joins. One calcium (Ca–O = 2.41 Å) is exactly symmetric with re-

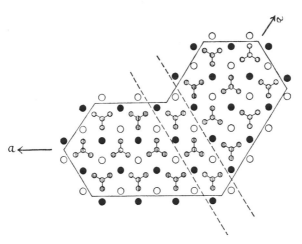

Fig. 28. The (110) twinning of aragonite (from BRAGG and CLARINGBULL). The figure shows two calcium layers and the interleaved CO₃ layer (corrugation not indicated). – If we lump together in two sets the short Ca–O distances (2.41 Å; 2.45 Å) and the long ones (2.53 Å; 2.54 Å; 2.66 Å) and disregard the differences within each set, then the structure between the broken lines, i.e. in the twin boundary, is consistent with the symmetry scheme in either individual

gard to the CO₃, the remaining two (Ca–O = 2.45 Å) approximately so (Fig. 27).

When we try to analyze the Ca–O bond lengths in relation to the position of the respective Ca with regard to the CO₃ groups, we find a long bond length (2.54 Å) for that Ca which is linked symmetrically to two oxygens of the same CO₃ group, i.e. aligned toward the interstice between the two oxygens. The shortest Ca–O (2.41 Å) marks the Ca which is linked strictly to one oxygen in the symmetrical position, i.e. which is farthest removed from the interstice. For the intermediate calcium atoms, the shortest characteristic Ca–O lengths (2.53 Å and 2.45 Å) decrease with increasing angular distance from the projection of the O–O interstice. This situation is reminiscent of a trend observed in the simple rhombohedral carbonates. In this series, the calcium in calcite shows the greatest cation-oxygen distance and it is located closest to the interstice between two oxygens of the same CO₃ group, although it does not at all approach the symmetric (or nearly so) alignment occurring in aragonite. Moreover, the magnesium in magnesite with its smaller bond length is farther removed from the interstice (Fig. 5b), and the other cations (e.g. Mn and Fe; not shown in Fig. 5b), in the respective carbonates, occupy intermediate angular positions in accordance with their sizes.

Although simple rhombohedral carbonates of the larger cations Sr, Pb, Ba are not available for comparison, the positions of the large cations in the double carbonates $PbMg(CO_3)_2$ and $BaMg(CO_3)_2$ (Fig. 17) may serve to confirm the rule that the tendency of the cation to approach the interstice, i.e. to be bonded to two oxygens of the same CO_3, increases with increasing cation-oxygen bond length; and vice versa. This behavior suggests that the gaps between two oxygens of CO_3 appearing in the current schematic representations of CO_3 (Fig. 5) are perhaps not real, at least not in determining the locations of bonded cations. At any rate, the rule is not amenable to simple electrostatic explanations based on the assumption of centered charges for the components of CO_3. It might be used rather to deduce information concerning the charge distribution within the CO_3 group. Meanwhile, even though purely empirical in character, the rule can help us to understand why aragonite-type structures are preferably formed by the carbonates of the larger cations (Sr, Pb, Ba). Only these, and Ca under certain conditions, are susceptible to being linked to two oxygens of the same CO_3 as required by the aragonite pattern. On the other hand, the marked tendency observed for the smaller cations (e.g. Mn, Fe, Mg) in their rhombohedral carbonates, to touch one oxygen of a CO_3 and to be far removed from another of the same CO_3, correlates with the absence of aragonite-type carbonates for these cations. Further use of the above deduced rule will be made on pp. 115—131 when we explain the formation of aragonite from solutions containing Mg^{2+} and try to elucidate the conditions for the radial and tangential orientations of the c axis of aragonite in the growth of oöids.

In connection with the latter task it is useful to point out another feature of the structure of aragonite in which it is distinct from that of calcite. Both structures are characterized by cation layers perpendicular to the c axes. In both these basal layers we find rows of calcium ions not separated by CO_3 groups. In calcite, the spacing of the calcium atoms (ions) in such basal strings is equal to $a_{hex} = 4.99$ Å, and in aragonite basal strings along $a = 4.96$ Å and parallel to the diagonal of the (a, b) outline with a spacing of 4.70 Å may be discerned. In addition, aragonite exhibits strings of Ca along c with the spacing of $c = 5.74$ Å. It is true that these strings are encroached upon by CO_3 groups, but calcium neighbors are not fully separated by CO_3, as seen in Fig. 27 where the Ca strings along c as well as the CO_3 strings running parallel are shown on end. In contrast, along the c axis of calcite, we find strings of alternating Ca and CO_3 (Fig. 3), the calcium atoms spaced at $\frac{c}{2} = 8.53$ Å thus being completely separated by CO_3 groups. In summary, the aragonite structure is featured by strings along c of Ca and CO_3, which correspond to similar strings of As and Ni in niccolite, whereas the cation rows along the c axis of the calcite-type structures are interrupted by an imbricate array of

CO_3 groups (Fig. 11). In both structures, we can, of course, find still other rows of Ca. However, these make oblique angles with the basal cation layers. For example, in calcite they run along the short diagonals of the cleavage faces (Fig. 2) with a spacing of 4.05 Å. It is obvious that such oblique rows are of no use in explaining fibrous growth in the direction of the c axes.

Results of the refined determination of the aragonite structure could be discussed and tabulated above, because Drs. A. DAL NEGRO and L. UNGARETTI, Pavia, Italy, kindly made available their unpublished manuscript, of which an abridged version appeared later in the American Mineralogist (DAL NEGRO and UNGARETTI, 1971). In the same issue, DE VILLIERS (1971) published his refinement of the aragonite structure. His results are identical with those of the Italian authors within the limits of error. DE VILLIERS also carried out refined determinations of the structures of strontianite, $SrCO_3$, and witherite, $BaCO_3$. These minerals, and also cerussite, $PbCO_3$, have long been known to be isomorphs of aragonite on the basis of crystal morphology and similar lattice constants (see Table 18).

DE VILLIERS found that strontianite and witherite are indeed isostructural with aragonite. In particular, the coordination features are identical and all bond lengths show the same trends as in aragonite, although the cation-oxygen distances become more uniform with increasing cation size and deviate somewhat less from the average bond length than in

Table 18. Lattice dimensions (in Å) for aragonite-type carbonates

		a	b	c	Source
Aragonite	$CaCO_3$	4.9616	7.9705	5.7394	DAL NEGRO et al.;
		4.9614	7.9671	5.7404	DE VILLIERS;
		4.959	7.968	5.741	
					SWANSON, FUYAT and UGRINIC (1954)
Strontianite	$SrCO_3$	5.107	8.414	6.029	
		5.090	8.358	5.997	DE VILLIERS;
Cerussite	$PbCO_3$	5.195	8.436	6.152	SWANSON and FUYAT
Witherite	$BaCO_3$	5.314	8.904	6.430	(1953)
		5.3126	8.8958	6.4284	DE VILLIERS;

The data of SWANSON et al. were determined by the diffractometer method at 26° C on powders of extremely pure synthetic materials. The values of DAL NEGRO and UNGARETTI are for a single aragonite crystal from material from Vertaizon-Alvernia containing less than 0.1% Sr. DE VILLIERS used minerals of similar purity, except for the strontianite which contains about 7 mol% Ca. For this reason the values for the strontianite are low in comparison to those determined by SWANSON et al. on purified $SrCO_3$.

5 Lippmann

aragonite. It is also remarkable that deviations from BRAGG's idealized structure become less pronounced with increasing size of the cation. The fact that, in contrast to $CaCO_3$, aragonite-type structures constitute the stable phases of $SrCO_3$ and $BaCO_3$ (see p. 99), appears thus to be correlated with greater structural symmetry.

Although the minerals strontianite and witherite are not of rock-forming importance, it would be desirable to discuss their structures in greater detail with special reference to aragonite. When the geochemical importance of the incorporation of strontium into aragonite is considered, interesting results may perhaps come from a specific comparison of aragonite and strontianite. However, due to the recent date of DE VILLIERS' publication, such an analysis is not available.

III. Miscellaneous Carbonates

The minerals whose structures have been discussed so far include the most important carbonates, which account for the composition of the overwhelming majority of all carbonate sediments, both ancient and recent. However, many more carbonate minerals have been reported from the sedimentary cycle. Although most of these appear to be restricted to very special conditions and to specific environments, some of them may be of more wide-spread occurence, since it is likely that they have frequently escaped detection on account of their ephemeral character, due to poor stabilities in general, or high solubilities in particular. The rarer carbonate minerals are introduced here not only for the sake of completeness, but also because some of them may have played a rôle in the formation of certain occurrences of the more common carbonate minerals, in part as precursors, in part as natural chemical reagents causing the precipitation and/or transformation of such minerals. If we regard vaterite below as a special case, the subdivision in hydrous carbonates (p. 71) and alkali bearing carbonates (p. 91), arbitrary though from the point of view of mineral systematics, may be taken to reflect something of these two possible petrogenetic rôles of the rarer carbonate minerals.

1. Vaterite

Vaterite, the third modification of $CaCO_3$, was named by W. MEIGEN (1911) in honor of H. VATER. This author seems to have synthesized the phase as early as 1894 and, relying mainly on the density, which he found lower than that of calcite, he suggested later that his phase might be different from both calcite and aragonite (VATER, 1902). JOHNSTON, MERWIN and WILLIAMSON (1916) seem to have prepared the same material, for which they determined the refractive indices $n_\omega = 1.55$ and $n_\varepsilon = 1.65$, and they referred to it as μ–$CaCO_3$. X-ray powder diagrams of VATER's modification of $CaCO_3$ published by RINNE (1924), HEIDE (1924) and v. OLSHAUSEN (1925) turned out in fact to be different from those for calcite and aragonite, thus confirming the phase as an additional $CaCO_3$ polymorph.

The first occurrence of vaterite in nature was reported by MAYER and WEINECK (1932), who asserted its presence in repair tissues of the fractured shells of certain gastropods. Although the X-ray evidence adduced by these authors appears scanty in the light of modern standards, similar results have been obtained later by WILBUR and WATABE (1963) (1960, quoted by KAMHI, 1963). PHEMISTER, ARONSOHN and PEPINSKY (1939) determined vaterite in gallstones along with calcite, aragonite and apatite. PRIEN and FRONDEL (1947) found no vaterite in human urinary calculi although they expressly considered the possibility. The status of vaterite as a mineral has been established by McCONNELL (1960) who observed it as a natural alteration product in hydrogel pseudomorphs after larnite, Ca_2SiO_4, from Ballycraigy, Larne, Northern Ireland. The formation of vaterite in hydrated portland cements evidently proceeds from similar materials, i.e. high-calcium silicates (COLE and KROONE, 1959; SCHRÖDER, 1962), but these occurrences are obviously not entirely natural in origin.

In spite of its frequent occurrence as a product of artificial precipitations of $CaCO_3$, vaterite has not been found in natural carbonate sediments so far. If we look for possible occurrences of vaterite in the sedimentary cycle we may expect its formation only under very specialized conditions, e.g. from certain saline waters in continental environments, and even there merely as a transitory phase. Judging from experiments by FLÖRKE and FLÖRKE (1961), vaterite might form in nature when waters rich in alkali carbonate react with gypsum. Although vaterite is of little petrological importance, at least according to present knowledge, its structure should nevertheless be of some interest in connection with the general subject of the polymorphism of $CaCO_3$.

Vaterite is optically positive, unlike calcite and aragonite. W. L. BRAGG (1924a, see BRAGG and CLARINGBULL) has explained the optically negative character of the latter by the coplanar arrangement of the CO_3 groups parallel to the basal plane of these structures. BUNN (1945, 1961) concluded from the optically positive character of vaterite that the planes of the CO_3 groups must be oriented perpendicular to the basal plane of the structure, i.e. parallel to the c axis. MEYER (1960) proposed a structural scheme for vaterite which is based on the niccolite pattern (Fig. 24) in a similar fashion as the aragonite structure. However, the substitution by Ca and CO_3 is thought to take place the opposite way in comparison with the derivation of the aragonite structure, i.e. the vaterite scheme is obtained by replacing Ca for Ni and CO_3 for As. On account of the shape of the CO_3 group and its orientation as deduced from the optical character, the a axes ($a = 7.15$ Å) have to be chosen $\sqrt{3}$ times the length of a' axes ($a' = 4.13$ Å) of the underlying niccolite cell (Fig. 30), in order to arrive at a unit cell compatible with hexagonal symmetry. This latter is indicated not only by crystal morphology and the optical properties but also by the X-ray diffraction data of vaterite (MEYER, 1969). Certain weak diffraction features observed by MEYER lead to a c spacing of 16.9 Å, which is doubled compared to previous determinations (v. OLSHAUSEN; McCONNELL; MEYER, 1960) and also with regard to the simple niccolite pattern. The systematic extinctions observed by MEYER (1969) indicate $P6_3/mmc$, $P6_3\,mc$, or $P\bar{6}2c$ as possible space groups. The diffuse character of certain X-ray reflections induced MEYER to evaluate models for disordered structures involving, among other, incomplete statistical occupations of atomic sites, as had been considered for vaterite also by KAMHI (1963). With regard to the manifold possibilities for disorder pointed out by MEYER (1969), the original paper must be consulted.

Table 19. Atomic sites in vaterite model for the sub-cell with $c' = \frac{c}{2} = 8.47$ Å and $a = 7.15$ Å, simplified from the structure proposed by BRADLEY, GRAF and ROTH (1966) for vaterite-type high-temperature $YbBO_3$.
Space group $P6_322 - D_6^6$ (No. 182; International Tables I)

	Wyckoff notation	Point symmetry	Coordinates
$2\,Ca_I$	b	32	$0,0,\frac{1}{4}$; $0,0,\frac{3}{4}$.
$2\,Ca_{II}$	c	32	$\frac{1}{3},\frac{2}{3},\frac{1}{4}$; $\frac{2}{3},\frac{1}{3},\frac{3}{4}$.
$2\,Ca_{III}$	d	32	$\frac{1}{3},\frac{2}{3},\frac{3}{4}$; $\frac{2}{3},\frac{1}{3},\frac{1}{4}$.
$6\,C$	g	2	$x,0,0$; $0,\ x,0$; $-x,-x,0$; $-x,0,\frac{1}{2}$; $0,-x,\frac{1}{2}$; $x,\ \ x,\frac{1}{2}$. $x = 0.423$
$6\,O_I$	g	2	$x,0,0$; etc. $x = 0.602$
$12\,O_{II}$	i	1	$x,\ \ y,\ \ z$; $-y, x-y,\ \ z$; $y-x,-x,\ \ z$; $y,\ \ x,\ -z$; $-x, y-x,\ -z$; $x-y,-y,-z$; $-x,-y,\frac{1}{2}+z$; $y,y-x,\frac{1}{2}+z$; $x-y,x,\frac{1}{2}+z$; $-y,-x,\frac{1}{2}-z$; $x,x-y,\frac{1}{2}-z$; $y-x,y,\frac{1}{2}-z$. $x=\frac{1}{3}$; $y=0$; $z=0.131$

The numerical parameters were chosen to yield uniform $Ca-O_{II}$ distances of 2.59 Å, and to conform with the CO_3 dimensions known from calcite. On account of $y = 0$ for O_{II}, this occupancy is also compatible with the space group $P\bar{6}2c - D_{3h}^4$ (No. 190) for the true (doubled) cell with $c = 16.94$ Å.

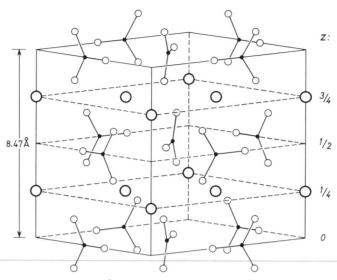

Fig. 29. Sub-cell ($c' = \frac{c}{2} = 8.47$ Å) of vaterite, $CaCO_3$, simplified from the structure of BRADLEY, GRAF and ROTH for high-temperature $YbBO_3$, and based on the data of Table 19. Note CO_3 groups standing vertical in the c direction. Of the z coordinates on the right side, $\frac{1}{4}$ and $\frac{3}{4}$ refer to Ca levels, 0 and $\frac{1}{2}$ to the centers of the CO_3 groups

For the sub-cell with $c' = \frac{c}{2} \simeq 8.5$ Å an idealized ordered vaterite structure may be developed starting from the pattern proposed by BRADLEY, GRAF and ROTH (1966) for the vaterite-type high-temperature modification of $YbBO_3$. The atomic coordinates according to the space group $P6_322$ are given in Table 19, and the model is shown in perspective in Fig. 29. The numerical value of $z = 0.131$ for the oxygen O_{II} conforms with the known dimensions of the CO_3 group. $y = 0$ and $x = \frac{1}{3}$ were chosen to obtain the perpendicular CO_3 orientation and a uniform bond distance of 2.59 Å from O_{II} to the different sets of calcium Ca_I, Ca_{II} and Ca_{III}. The Ca–O distance of 2.59 Å is close to the longer Ca–O occurring in aragonite (Table 16). However, the oxygens O_{II} surround the calcium atoms (ions) in an octahedral fashion reminiscent of the calcite structure (Fig. 30). The CO_3 dimension then fixes C and O_I. The latter surround the calcium atoms by way of additional coordination polyhedra characterized by considerably greater Ca–O lengths than the octahedra mentioned already. The coordination polyhedron around Ca_{II} and Ca_{III} has Ca–O_I = 3.05 Å and is intermediate between an elongate octahedron and a trigonal prism. Ca_I–O_I is 3.55 Å, and it cannot be decided to what extent the almost regular octahedron in question is essential for the

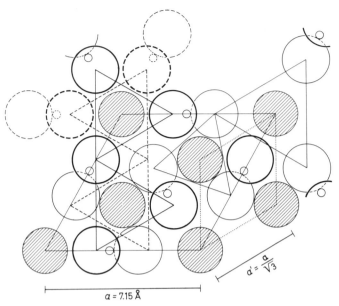

Fig. 30. Vaterite structure (simplified from BRADLEY GRAF, and ROTH; cf. Fig. 29) projected in the c direction. The Ca (shaded circles) belong to the level with $z = \frac{3}{4}$. The carbons (small circles) and the unique oxygens O_I (normal circles) are at $z = 1$. Two oxygens O_{II} (heavy circles), belonging to the same CO_3 and disposed symmetrically above and below the level with $z = 1$, overlap in this projection. Two CO_3 groups (broken circles) of the level at $z = \frac{1}{2}$ are shown outside the lozenge-shaped cell outline. Of the CaO_6 octahedra characterized by Ca–O = 2.59 Å, the basal triangles above (full lines) and below (broken lines) the Ca have been drawn. Of the coordination polyhedra formed by O_I, only the top basal triangles are outlined. The smaller of these belongs to the distorted prism (or octahedron) with $Ca_{II,III}$–O_I = 3.05 Å, whereas the large triangle marks the octahedron with Ca_I–O_I = 3.55 Å. It is seen that the niccolite-type sub-cell (dotted lines), with $a' = a/\sqrt{3}$, is devoid of trigonal axes, required for hexagonal symmetry, because of the shape of CO_3
(MEYER, 1960)

bonding of Ca_I. However, the proposed structural scheme is still susceptible to refinement, which has not yet been carried out in the instance of $CaCO_3$. Possible variations include deformations of the CO_3 group, shifting it along the a directions, and tilting it (slightly) from its perpendicular alignment.

Such a refinement for the space group $P6_322$ is necessarily restricted to the sub-cell and disregards the fact that the true cell of vaterite has a doubled c dimension and is characterized by space groups different from $P6_322$. From this point of view, it is remarkable that a large cell composed of two sub-cells adjacent in the c direction may be constructed which complies with the symmetry $P\bar{6}2c$, one of the possible space groups as determined by MEYER (1969). This is seen by transforming the coordinates of two adjacent small cells to those of one composite cell (not actually carried out here). Two cells of the symmetry $P6_322$ will have the composite symmetry $P\bar{6}2c$ only for the occupancy given in Table 19 and, in particular, for strictly perpendicular CO_3 groups, i.e. with $y=0$ for O_{II}. However, these restrictions may be dropped as soon as the occupancy of the large cell is established, and adjustments along the lines stated above (deformations, shifts and slight tilts of CO_3) become then possible, but this time according to the degrees of freedom of $P\bar{6}2c$. It is not likely that such a refinement will shift the short $Ca-O_{II}$ lengths appreciably out of the range the $Ca-O$ distances in aragonite. However, a variety of different, but not greatly deviating $Ca-O$ distances will evidently be the result. Misalignments of the CO_3 groups with respect to these distances, among them tilts of the slightly inclined CO_3 in the "wrong" sense, will then yield a great many possibilities for disorder, and these offer alternative explanations for the diffuse X-ray diffraction features observed by MEYER (1969). Whatever structural model will finally turn out to be the right one, the lattice constants of vaterite (MEYER's data, Table 19) and the occupancy of the sub-cell by 6 $CaCO_3$ lead to the density of 2.65_8 in any case, and thus require a rather loosely packed structure, which is in harmony with the poor stability of the mineral.

2. Hydrous Carbonates and the Magnesite Problem

a) Hydrous Magnesium Carbonates

Among the cations forming the carbonate minerals of sedimentary rocks, magnesium holds the second place after calcium. Although most of the magnesium of sedimentary carbonates is contained in dolomite, and some also in magnesite, both of which are typically anhydrous minerals, magnesium shows a marked tendency to form hydrous carbonate minerals. These are:

hydromagnesite	$Mg_5(CO_3)_4(OH)_2 \cdot 4H_2O$ or
	$Mg_4(CO_3)_3(OH)_2 \cdot 3H_2O$
artinite	$Mg_2(CO_3) (OH)_2 \cdot 3H_2O$
nesquehonite	$MgCO_3 \cdot 3H_2O$
lansfordite	$MgCO_3 \cdot 5H_2O$

Barringtonite, $MgCO_3 \cdot 2H_2O$ (NASHAR, 1965) and dypingite, $Mg_5(CO_3)_4(OH)_2 \cdot 5(or 6)H_2O$ (RAADE, 1970) are known so far only from one locality each. Two formulae, difficult to distinguish by chemical analysis, are quoted from the literature for hydromagnesite. The second formula will be used in the following, chiefly because it was adopted by LANGMUIR (1965) in his thermodynamic study on magnesium carbonates. There is nevertheless serious analytical evidence in favor of the first formula (see BALCONI and GIUSEPPETTI, 1959).

The hydrous carbonates of magnesium commonly occur as alteration products of rocks rich in magnesium, notably serpentines (for typical localities see PALACHE, BERMAN and FRONDEL). However, the first occurrence of nesquehonite and lansfordite was reported from a coal mine (at Nesquehoning near Lansford, Pennsylvania) and thus obviously belongs to the sedimentary cycle. More recently, hydromagnesite has been found in the recent carbonate sediments of certain ephemeral lagoonal lakes of South Australia, otherwise known for the formation of recent dolomite. Judging from the descriptions of this occurrence by VON DER BORCH (1965) and ALDERMAN (1965), hydromagnesite, together with aragonite, appears to play the rôle of a precursor of dolomite and magnesite. ALDERMAN (1965) obtained artificial hydromagnesite, in addition to aragonite, by adding sodium carbonate to sea-water. More generally, room-temperature experiments aiming at the precipitation of magnesite and dolomite have commonly yielded products containing hydromagnesite and/or nesquehonite.

It is not possible here to review the numerous publications dealing with the preparation of these phases. From these publications, most of which are found quoted in a paper by LANGMUIR (1965), certain generalizations may be abstracted which have in part been verified by the present writer (LIPPMANN, unpublished).

A phase yielding the X-ray powder pattern of hydromagnesite is usually produced when solutions of magnesium salt (e.g. magnesium chloride or sulphate) and alkali carbonates are mixed at such concentrations as will lead to instantaneous precipitation. Although compositional deviations from the two above quoted hydromagnesite formulae have been reported for such precipitates, involving not only the water content but also the ratio of hydroxide to carbonate, there can be no doubt as to the general identity of instantaneously precipitated basic magnesium carbonates with hydromagnesite. The exact mechanisms for the compositional variability are not yet clear, but crystal disorder and superficial adsorption by the fine-grained precipitates may be suggested as explanations. At any rate, the other basic magnesium carbonate, artinite, has not yet been observed among such laboratory preparations, and the formation of a similarly composed material has been claimed so far in one instance (KAZAKOV et al., quoted by LANGMUIR). The concentrations required to precipitate basic magnesium carbonate decrease as the temperature is raised from normal toward 100° C. This is mainly due the tendency of the HCO_3^- produced by the precipitation (see below) to show an increasing decomposition pressure of CO_2 with rising temperature.

If a precipitate obtained at room-temperature is left in contact with its mother liquor, nesquehonite is commonly found to have crystallized after some days, often in the form of radial-fibrous spherules. Obviously the OH^- required in the primary precipitation of hydromagnesite is available from the hydrolysis of the CO_3^{2-}, and HCO_3^- is formed instead:

$$CO_3^{2-} + H_2O \rightarrow HCO_3^- + OH^-. \tag{1}$$

The formation of hydromagnesite then proceeds according to:

$$4\,Mg^{2+} + 5\,CO_3^{2-} + 5\,H_2O \rightarrow Mg_4(CO_3)_3(OH)_2 \cdot 3\,H_2O + 2\,HCO_3^-. \tag{2}$$

The subsequent slow transformation to nesquehonite

$$\begin{aligned} 3\,Mg_4(CO_3)_3(OH)_2 \cdot 3\,H_2O + 6\,HCO_3^- + 21\,H_2O \\ \rightarrow 12\,MgCO_3 \cdot 3\,H_2O + 3\,CO_3^{2-} \end{aligned} \tag{3}$$

takes place, because by reaction (2) the solution has become depleted of CO_3^{2-}, whereas HCO_3^- has been added by the same reaction. Consequently, according to

$$\frac{[HCO_3^-] \cdot [OH^-]}{[CO_3^{2-}]} = k_{hydr.} \tag{4}$$

the concentration of OH^- has decreased to such a degree that hydromagnesite becomes unstable with respect to nesquehonite. It may be concluded that hydromagnesite and nesquehonite are mutually in equilibrium under these conditions, although the primary instantaneous formation of hydromagnesite and its dilatory transformation to nesquehonite indicate that the crystallization of the latter is delayed by kinetic inhibitions. It is conceivable that without these inhibitions hydromagnesite would not appear at all as an intermediate phase in precipitations with alkali carbonate, but would form only when both alkali carbonate and hydroxide are added to a solution of magnesium salt.

Nesquehonite crystallizes as a primary phase below about 60° C when CO_2 escapes slowly from solutions of magnesium bicarbonate, or from mixed solutions of magnesium salt and alkali bicarbonate, according to:

$$Mg^{2+} + 2\,HCO_3^- + 2\,H_2O \rightarrow MgCO_3 \cdot 3\,H_2O + CO_2. \tag{5}$$

Lansfordite forms under similar conditions, but below about 10° C.

Because of its preferred formation in the presence of bicarbonate, nesquehonite has sometimes been interpreted as a basic bicarbonate. Recently MORANDI (1969) wrote its composition as

$$MgCO_3 \cdot Mg(HCO_3)(OH) \cdot 5H_2O .$$

It is, however, highly unlikely that two ions which would react according to (4) can be present in the same crystal structure. The crystallization of nesquehonite from solutions rich in HCO_3^- is adequately explained by the suppression, according to (4), of the OH^- concentration below the level required for the precipitation of hydromagnesite according to the solubility product:

$$[Mg^{2+}]^4 \cdot [CO_3^{2-}]^3 \cdot [OH^-]^2 = k_{sH} .$$

For the constant k_{sH} LANGMUIR has proposed a value of about 10^{-30}. For the solubility product of nesquehonite

$$[Mg^{2+}] \cdot [CO_3^{2-}] = k_{sN}$$

various authors agree that the constant k_{sN} is of the order of 10^{-5}.

It is extremely interesting that this is considerably greater, by several orders of magnitude, than most of the values deduced by LANGMUIR from the experimental data of various authors for the solubility product of magnesite

$$[Mg^{2+}] \cdot [CO_3^{2-}] = k_{sM} .$$

The wide range of these values, from 10^{-7} to 10^{-10}, suggests that reliable determination of the solubility product for magnesite from solubility data is perhaps impossible. From the fact that magnesite is isostructural with calcite and from the similarity of the crystal ("lattice") energies (see p. 15), one may conclude that the solubilities of both minerals, should be of the same order of magnitude. Such a conclusion is borne out by the values $k_{sM} = 10^{-8.1}$ and $k_{sC} = 10^{-8.4}$ which LANGMUIR deduced from thermochemical data. The value k_{sC} for calcite is practically the same as the experimental solubility product $10^{-8.34}$ of GARRELS, THOMPSON and SIEVER (1960). This agreement may be taken to substantiate $k_{sM} = 10^{-8.1}$ for magnesite, since it is based on thermochemical data of the same quality as those for calcite. The magnitude of $k_{sN} \simeq 10^{-5}$ for nesquehonite then inevitably leads to the conclusion that nesquehonite, and also hydromagnesite, which obviously exists in equilibrium with the former, are metastable with respect to magnesite. Such a view may appear unusual at first sight, since we are accustomed to take it for granted that reactions involving ions in solution proceed practically instantaneously in the direction determined by thermodynamic properties. This attitude lead LANGMUIR to dispute those thermodynamic data for magnesite which are independent of direct solubility determinations and which yield $k_{sM} = 10^{-8.1}$. Of the experimental solubility data for magnesite, he preferred those closest to the solubility of nesquehonite, but for which the character of the solid phase was least authenticated. This way he secured stability fields for both hydromagnesite and nesquehonite, and assumed the lowest temperature for which the synthesis of magnesite has been reported, of about 60° C, as the transformation temperature for the equilibrium magnesite–nesquehonite.

We are thus left with the choice either to accept this latter view of LANGMUIR's or to assume that nesquehonite and hydromagnesite are metastable down to at least 25° C and that magnesite, on the basis of its thermodynamic properties, is the only stable magnesium carbonate in this temperature range. Both views imply serious kinetic complications. For the first view we have to assume that the dissolution of magnesite is inhibited, since in all solubility determinations for well defined magnesite the mineral has proved to be considerably less soluble than nesquehonite. For the second view kinetic inhibitions must be assumed which preclude the crystallization of magnesite near room temperature.

This "magnesite problem" is not the only case where reactions which should proceed according to thermodynamics fail to do so because they are inhibited by inherent activation barriers which are practically insurmountable at the rather low thermal energies available near room temperature. The most familiar example is that all organic matter, living and dead, is thermodynamically unstable in contact with the oxygen of the air, and so life is possible only due to a great number of kinetic inhibitions, which make the oxidation of organic substances at ordinary temperature a rather circuitous affair, requiring the cooperation of a great number of enzymes. As for the mineral kingdom, most of our knowledge concerning the petrography of igneous and metamorphic rocks is due to the fact that phase changes required by thermodynamics, e.g. weathering and adjustments to different metamorphic facies, are extremely sluggish or even completely inhibited by kinetic obstacles. In many such instances thermodynamic properties have furnished valuable information concerning the stability of mineral phases before direct experimental determinations of equilibrium conditions became possible. The most prominent example is the system graphite – diamond (see e.g. CORRENS, p. 187). In view of this and other examples we should give more weight to the solubility product of magnesite $k_{sM} = 10^{-8.1}$ as deduced from thermodynamic properties than to the many discordant experimental solubility data. At any rate, we should investigate which of the two reactions, the crystallization or the dissolution of magnesite, is more liable to kinetic inhibition.

For this purpose we ought to know the crystal structure not only of magnesite but also of hydromagnesite and nesquehonite. However, the two latter structures have not yet been analyzed. Nevertheless, we may infer certain aspects of these structures, very general though in character, from the known structures of other hydrous magnesium minerals. The structure of artinite, $Mg_2(CO_3)(OH)_2 \cdot 3H_2O$, has been analyzed by JAGODZINSKI (1965). It contains the magnesium in octahedral coordination by oxygens. In every octahedron, three of the oxygens are from hydroxyls (OH) and two from water (H_2O). The remaining sixth octahedral corner is statistically occupied by oxygen either from the CO_3 group or from an H_2O molecule. The MgO_6 octahedra are linked by common OH–OH and H_2O–HO edges and form chains along the b axis. Among themselves, the chains are connected mainly by hydrogen bonds extending from H_2O to CO_3. Generally speaking, the magnesium in artinite is immediately surrounded mainly by hydroxyls and water molecules.

We may now reasonably infer that hydromagnesite contains MgO_6 octahedra as well. Although we know nothing of the mutual articulation of these octahedra, it is safe to assume from the chemical similarity with artinite that part of the oxygens directly bonded to magnesium are contributed by OH and H_2O, the remainder by CO_3. At least part of the CO_3 oxygens, however, will be linked to magnesium indirectly via water molecules by way of hydrogen bonds as in artinite.

The tendency of the water molecules to be bonded directly to the cation is observed in most hydrated salts. Magnesium in particular has as many immediate H_2O-oxygen neighbors as are available for a given composition. Of course, the usual octahedral coordination of magnesium limits the maximum number of water neighbors to six. This occurs in epsomite, $MgSO_4 \cdot 7H_2O$, where $(H_2O)_6Mg$ octahedra are held together by SO_4 tetrahedra by way of hydrogen bonds, and in part also via the seventh H_2O molecule (BAUR, 1964). In leonhardtite, $MgSO_4 \cdot 4H_2O$, (BAUR, 1962) and bloedite (astrakhanite), $Na_2Mg(SO_4)_2 \cdot 4H_2O$, (RUMANOVA and MALITSKAYA, 1960) four oxygens in the MgO_6 octahedra are contributed by H_2O, the remaining two by SO_4 groups. On the basis of this experience, we may infer that the structure of nesquehonite, $MgCO_3 \cdot 3H_2O$, contains MgO_6 octahedra as well, where three corners are occupied by H_2O and three by oxygens from three different CO_3 groups.

b) Cation Hydration and the Dehydration Barrier

According to what we have deduced from the known structures of hydrated magnesium salts, the configurations of hydromagnesite and nesquehonite will most probably not deviate from the octahedral coordination of oxygen about magnesium as in magnesite. From this point of view, the crystallization and the dissolution should proceed with the same ease or the same difficulty, and thus at comparable rates for all three magnesium carbonates. It cannot be seen why magnesite should react in a much more sluggish fashion in aqueous solutions than do the hydrous carbonates. The only fundamental difference, which might be of use in explaining the singular kinetic behavior of magnesite, may be seen in the fact that it is anhydrous, whereas hydromagnesite and nesquehonite contain water molecules bonded to magnesium. If we consider that ions are hydrated in solution, i.e. bonded by molecules of water, then the condition of the magnesium in the hydrated carbonates is akin, to some extent, to that in the dissolved state. On the other hand the transition, both ways, between solution and anhydrous magnesite involves a more fundamental change of the state of the magnesium.

In order to appreciate the importance of this change, a short account of the hydration of ions will be given in the following paragraphs. For more detailed treatments of the subject, texts of physical chemistry may be consulted (e.g. HARVEY and PORTER, especially section 9—2). Owing to the partially ionic character of the bonding within the water molecule, it is endowed with a strong dipole moment of 1.84 debye. This value is exceeded among the inorganic compounds only by that ot HF (1.98 debye), and by a number of organic molecules, and is reflected by the high dielectric constant of water (of 81). The water molecule therefore shows strong electrostatic interactions with all kinds of ions. The negative charge center of the dipole nearly coincides with the center of the oxygen, which may be regarded as an O^{2-} ion whose charge is balanced by two H^+ ions or protons. These are about 1 Å distant from the oxygen center, with which they from an angle of 104.5°, close to the angle formed by the threefold axes of the regular tetrahedron. The direction of the

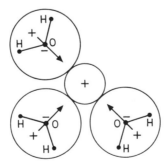

Fig. 31. Water dipoles linked to a positively charged ion. The approximate charge centers of the dipoles are indicated. In the present two-dimensional sketch only part of the water molecules which would surround a dissolved cation in a three-dimensional arrangement, have been shown. (From HARVEY and PORTER)

dipole moment is determined by the bisectrix of the H–O–H angle. The electrostatically most stable arrangement of the water molecule, with respect to a positively charged cation, obtains when the dipole direction points at the center of the cation (Fig. 31). The electrostatic energy for the ion-dipole bond is given by

$$E = -\frac{e \cdot \mu}{r^2} \cdot \cos(\mu, r) \quad \text{(see Joos, 1934)} \quad (6)$$

where e is the charge of the ion, μ the dipole moment, r the distance from the cation center to the center of the negative charge of the dipole (the oxygen center for H_2O), and (μ, r) the angle formed by the dipole direction μ and the join cation-oxygen r, (μ, r) being zero for the configuration shown in Fig. 31. It may be noted that the absolute value of the ion-dipole energy declines rather rapidly, with the reciprocal of the square of the distance r, in contrast with the energy of an ion-ion bond, which is a function of the reciprocal of r itself, according to Coulomb's law. The ion-dipole bond is thus less influenced by additional ligands bonded to one same ion than the ion-ion bond. Nevertheless, the electrostatic calculation, according to (6), of the energy of hydration would be prohibitive for a dissolved ion, since the number of water molecules bonded directly and indirectly (via directly linked H_2O) can only be roughly estimated. The hydration number for bivalent cations is differently estimated at from 9 to 17. Moreover, deviations from simple ion-dipole models must be expected from the influence of bond contributions other than electrostatic. Therefore, the energies (enthalpies) for the complete dehydration of dissolved ions are better deduced from thermochemical data. Such an analysis has been presented by Noyes (1962), who also discussed previous work (cf. Harvey and Porter). Numerical values for the enthalpies characterizing the hydration of bivalent cations are quoted in Table 20, column 2. The values, in the order of hundreds of kcal/mole, are similar in magnitude to crystal ("lattice") energies. In fact, the energy balance for the dissolution of an anhydrous crystal or for its crystallization from solution, i.e. its enthalpy of solution, is roughly equivalent to the difference between the hydration enthalpy for all components and the crystal energy. It is seen in Table 20 that the enthalpy of hydration increases markedly with decreasing ionic size. The hydration enthalpy of CO_3^{2-} ions has not yet been evaluated, but it will be certainly below 300 kcal/mole by virtue of the greater effective size of CO_3^{2-} relative to the S^{2-} ion, for which the value is 313 kcal/mole.

It is now tempting to look upon the enthalpies of hydration, preferably the rather large values for the bivalent cations, as barriers inhibiting crystallization. The enthalpy of hydration has, in fact, to be put into the system when a dissolved ion, e.g. magnesium, is to be incorporated in an

Table 20. Energies of interaction for H₂O dipoles with ions

1 X	2 Enthalpy of hydration ΔH^* in kcal/mole	3 X–O distance $(=r)$ sum of GOLDSCHMIDT ionic radii (in Å)	4 Energy of interaction of one H_2O with cation, according to (6) in kcal/mole
$(CO_3^{2-}$	$\gtrsim 300)$	—	—
$(S^{2-}$	313)	—	—
Ba^{2+}	326	2.75	33.6
Pb^{2+}	368	2.64	36.5
Sr^{2+}	354	2.59	37.9
Ca^{2+}	395	2.38	44.9
Mn^{2+}	455	2.23	51.2
Fe^{2+}	473	2.15	55.0
Mg^{2+}	473	2.10	57.7

Column 2: values from NOYES (1962) (cf. HARVEY and PORTER).
Column 4: The values substituted in (6) are: r according to column 3, and the dipole moment for water $\mu = 1.84$ debye.
The deviations of the values in column 2 for Fe, Mn, and especially Pb, from a consistent relationship with the distances in column 3 are perhaps indicative of the more marked tendency of these cations toward non-polar bonding.

anhydrous crystal, as e.g. magnesite. The enthalpy of hydration is about 20% greater for Mg^{2+} than for Ca^{2+}, whereas the crystal energy gained in crystallization is only 10% greater for magnesite (771 kcal/mole) than for calcite (701 kcal/mole).

This disparity enhances the impression that the high enthalpy of hydration for Mg^{2+} ion has something to do with the failure of magnesite to crystallize from aqueous solutions at ordinary temperature, in contrast to the ease with which calcite forms under comparable conditions. Such reasoning will, however, not lead to fundamentally different solubilities for both minerals. For if we add to the hydration enthalpies of Mg^{2+} and Ca^{2+} any value between 200 and 300 kcal/mole for the hydration of the CO_3^{2-} ion, the difference between both sums will amount to about 10%, the same as the difference between the crystal energies for magnesite and calcite. This is in harmony with the conclusion deduced above from the similarity of the crystal structures, viz. that the solubility of both minerals should be of the same order of magnitude.

Different behavior in crystallization may be predicted when the incorporation of the cation into the structure is viewed as a separate process, which it actually is, in the case of the growth of an ionic crystal from an ionized solution. It is sound to assume that the energy liberated on the entrance of a cation into the crystal structure amounts to half the crystal energy. This is strictly true for the simpler binary ionic crystals of NaCl type (see CORRENS, p. 167), but it should also apply to calcite-type carbonates with fair approximation, by virtue of their kinship with the NaCl type. From this point of view, half the lattice energy has to be subtracted from the cation hydration enthalpies, and the incorporation of the cations Mg^{2+} or Ca^{2+} is an endothermic process requiring 88 or 45 kcal/mole, respectively. Hence the growth of magnesite should be more strongly inhibited, by the dehydration barrier, than that of calcite, in harmony with laboratory experience. The energy budget will be about balanced by the subsequent entrance of a CO_3^{2-} ion. This process is exothermic (half the lattice energy minus the hydration enthalpy for CO_3^{2-}), and more energy is liberated in the case of magnesite, as its crystal energy is greater than that of calcite.

However, the foregoing explanation is not quite realistic, in that it implies that an ion is dehydrated completely in one step on entering a crystal structure. An ion, in order to be incorporated into a crystal, must first be absorbed by its surface. In this act, it loses part of the adhering water molecules. Others are retained and extend outward into the surrounding solution, or have one of their positively charged hydrogen ends linked, by way of hydrogen bonding, to an anion of the crystal surface. No difficulties can be seen for the adsorption of a cation. The energy to be put in for its partial dehydration will certainly not exceed the amount gained, about half the crystal energy. One might object that additional energy is needed for the complete dehydration of the neighboring CO_3 groups, already part of the crystal but still encumbered with water dipoles extending outward. However, the CO_3 group is relatively large in dimension, and it is unlikely that its free surface is completely occupied by water molecules. Some cation, perhaps incident from a favorable direction, will nonetheless succeed in becoming bonded to CO_3 and so be absorbed by the crystal surface.

The case is different for the cations adsorbed on the crystal surface. Because of their smaller dimensions, they will be more or less completely shielded by adhering water dipoles extending outward. Incoming CO_3^{2-} ions perhaps will not succeed in becoming adsorbed, because they cannot approach the surface cations close enough, because of the water molecules attached to the latter. The CO_3^{2-} ion is thus not sufficiently attracted to be able to supplant the water molecules.

At this stage of the discussion of the growth mechanism of ionic crystals from solution, information is needed concerning the strength of the bond between the water dipoles and the surface cations. Experimental data which are immediately applicable to our problem do not seem to be available, and it is an open question whether heats of desorption for water molecules are at all accessible to direct measurement for the conditions outlined above. It is, however, possible to calculate the interaction energy for an ion and one dipole according to (6), i.e. the energy to be expended for separating one particle from the other to infinite distance. For this purpose, in addition to the charge of the cation ($2e$) and the dipole moment of the water molecule (1.84 debye), the distance r from the center of the cation to the negative center of H_2O has to be substituted in (6). In the known crystal structures of hydrated magnesium salts, the H_2O–Mg distances show a certain spread, e.g. from 2.10 to 2.05 Å in $MgSO_4 \cdot 7H_2O$ (Baur, 1964). Therefore, in order to obtain standard values for ion-dipole interaction energies, the sum of the Goldschmidt ionic radii for oxygen and the cation was substituted for r in (6). The interaction energies thus calculated are listed in Table 20, column 4. They are probably low, especially for the small cations, because the H_2O–Mg lengths in crystals are smaller, on the average, than the sum of the ionic radii of Mg and O (see above). Since the values imply infinite separation of both partners, they are valid strictly for a gas phase expanded to infinite volume. Visualizing that an ion in solution, dehydrated already except for one water molecule, is finally stripped of the last molecule, we have to diminish the values in column 4, Table 20, by the heat of evaporation for water of 9.7 kcal/mole, since the water molecule in question ends up in the liquid phase. The relative difference between the interaction energies for small and large cations, remarkable already in column 4, Table 20, then becomes still more pronounced (Table 21, column 1). However, we are not interested actually in the removal of the last water dipole from a lone cation, but from a cation adsorbed on a crystal surface. In the latter case, the energy is certainly lower than in the former, because adjacent CO_3 groups, by virtue of their negative charge, counteract the attraction of the water dipole by the

Table 21. Tentative activation energies (in kcal/mole) derived from the ion-dipole interaction energies of Table 20, column 4:

	1 By subtracting the heat of evaporation for water (9.7 kcal/mole)	2 By subtracting the energy of the H_2O–Ba^{2+} bond
Ba^{2+}	23.9	0.0
Pb^{2+}	26.8	2.9
Sr^{2+}	28.2	4.3
Ca^{2+}	35.2	11.3
Mn^{2+}	41.5	17.6
Fe^{2+}	45.3	21.4
Mg^{2+}	48.0	24.1

positively charged cation. The strength of the bond between the negative oxygen end of the dipole and a surface cation is lowered by the negative charges of the neighboring near-surface CO_3, even in the case that one of the proton ends of the same dipole is linked to one of these CO_3.

We may now be tempted to calculate in detail the electrostatic bond energies for a water molecule forming different configurations with the ions on a growth face, say the cleavage rhombohedron of various rhombohedral carbonates, assuming that the constitution of such a face is the same as that known from structure analysis for its counterparts inside the crystal. However, although we can be reasonably sure that such an assumption is legitimate in its qualitative aspects, we must expect that the interatomic distances near the surface deviate from the ones known for the interior of the crystal. In particular, we must expect that the surface structure, notably the alignment of CO_3, is disturbed by the same water molecules for which we want to determine bond energies. If all these complications could be accounted for, and if the calculation could actually be carried out, the resulting energies for the bond between a surface cation and a water dipole would be lower, but of about the same order of magnitude as the interaction energies for a lone cation and a dipole. This is because the steep decline of the dipole potential with distance [cf. (6)] limits the influence of additional ligands of the cation in question. Moreover, we may also expect that the absolute differences in interaction energy between cation species will be less affected by additional neighbors of the cation than the absolute values themselves.

It is now worth noting, that the energies of interaction between a lone ion and a dipole as given in Table 21, column 1, conform, in their order of magnitude of tens of kcal/mole, with the activation energies known for other inorganic reactions characterized by slow rates at ordinary temperature or by complete inhibition (see texts of physical chemistry, e.g. HARVEY and PORTER). Bearing in mind that they are probably somewhat high, we may now tentatively accept the interaction energies as potential-barriers inhibiting the growth of anhydrous carbonates from aqueous solutions, and treat them as activation energies E_a determining the rates of crystallization. In doing so, we will not obtain any information concerning absolute rates, but growth rates of isostructural crystals may be compared. For these, all rate determining factors should be about the same, except for the Arrhenius factor, $\exp(-E_a/RT)$. By substituting the difference of the interaction energies of Ca^{2+} and Mg^{2+}, i.e. 12.8 kcal/mole, for E_a, we find that calcite should crystallize about 10^{10} times faster than magnesite at 25° C. Since the crystallization of calcite is by no means instantaneous, but is characterized by reaction times at least of the order of minutes to hours, it follows that magnesite, although thermodynamically stable, has such an extremely small growth rate at 25° C that it practically does not form. It is not possible to make up for this infinitesimal rate by a correspondingly high supersaturation, because this possibility is limited by the formation of hydromagnesite and nesquehonite. The growth of these hydrous carbonates is not complicated, in contrast to magnesite, by the reluctance of the magnesium ion to lose its last H_2O ligand(s) as these are incorporated into the crystal structures. Hence the hydrous carbonates precipitate or crystallize when their solubility products are exceeded, although they are more soluble than magnesite.

This state of affairs is not essentially changed by assuming a smaller value for E_a, say 10 kcal/mole. In this case a factor of about 10^8 is the result. A factor of this order would apply to the growth rate ratio of calcite versus siderite (10.1 kcal/mole). The experience that the latter, unlike magnesite, may be obtained artificially at ordinary temperature, most probably has something to do with the non-existence of a hydrous carbonate of iron, an empirical fact which still awaits explanation. Thus in contrast to the case of magnesium, extremely high supersaturations may be maintained for siderite, so that it finally forms, although in an extremely sluggish fashion and with poor crystallinity. On a qualitative scale, the ease of crystallization of rhodochrosite is intermediate between calcite and siderite, in harmony with the intermediate value of the interaction energy for the H_2O–Mn bond. At the same time, rhodochrosite precipitated from solutions of e.g. $MnCl_2$ and K_2CO_3 is more poorly organized than calcite obtained under analogous conditions, as indicated by the diffuseness of X-ray patterns (GOLDSMITH and GRAF, 1956). This appears to be related to water molecules contained in the structure in a random fashion.

So far, we have assumed tacitly that the formation of anhydrous carbonates is determined mainly by growth rates, and we have left out of consideration nucleation problems. It is easy to verify that the addition of magnesite nuclei is of no help in attempts to prepare magnesite at ordinary temperature. Moreover, given the close structural similarity, it would be difficult to understand why calcite does form spontaneous nuclei whereas magnesite apparently does not. In order to summarize the above explanation for the inhibited growth of magnesite by the reluctance of the Mg^{2+} ion to dehydrate and in order to give the interpretation a more graphic expression, we may contemplate the fate of a nucleus immersed in a solution supersaturated with respect to magnesite. Regardless of whether the nucleus was added on purpose or has formed spontaneously, the ions exposed at the surface, in particular Mg^{2+}, are occupied by water dipoles. Incoming hydrated cations, such as A in Fig. 32, become bonded by anions of the crystal surface, since these, on account of their size, are rather incompletely covered by water mole-

6*

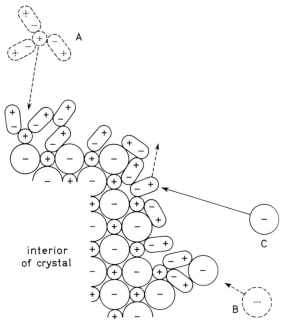

Fig. 32. A nucleus of magnesite, showing water dipoles bonded preferably to the cations of the crystal surface. For the sake of simplicity, the magnesite crystal is represented by an NaCl type structure. In order to emphasize the directional character of ion-dipole bonds, the water dipoles are represented as elongate particles, although their shape corresponds more closely to spheres. Moreover, it must be borne in mind that the present sketch fails to show any three-dimensional aspects of the hydration of the crystal surface. The dipoles making up the structure of the surrounding liquid have been omitted. The hydrated cation A and the anion B are shown dashed in their positions prior to the adsorption on the crystal surface

cules. Although, in this process, the cation loses part of the adhering water dipoles, other dipoles remain bonded and shield the cation toward the outside. Incoming anions, such as B, will not normally succeed in becoming bonded directly to surface cations, as would be required for an anhydrous crystal to grow. Instead they are weakly linked indirectly via water molecules, from which position they are easily released again into solution, unless the supersaturation is so high that the growth of a hydrous salt is possible. Only very few anions, such as C in Fig. 32, are endowed with enough energy (translatory or vibrational) that they are able, on hitting the crystal surface, to knock off the water dipole(s) from a surface cation and to become directly bonded to the latter. The number of these privileged or activated anions is determined mainly by an

Arrhenius factor $\exp(-E_a/RT)$ and is, of course, proportional to the total concentration of CO_3^{2-}. The influence of the Arrhenius factor explains why magnesite may be prepared at higher temperatures, near 200° C (MARC and ŠIMEK, 1913; JANTSCH and ZEMEK, 1949), although it is stable also at room temperature.

Magnesites that have undoubtedly formed at earth-surface temperatures have been described by GRAF, EARDLEY and SHIMP (1961) from the sediments of Glacial Lake Bonneville, and by VON DER BORCH (1965) and ALDERMAN (1965) from ephemeral coastal lakes in South Australia, where magnesite obviously develops from hydromagnesite. For the latter occurrence the authors have pointed out an influx of ground water unusually high in alkali carbonate and bicarbonate, and such an influence cannot be excluded for the sediments of Lake Bonneville. A high carbonate concentration entails a relatively high amount also of activated CO_3^{2-}, capable of breaking the hydration envelope of a surface Mg^{2+}. It is thus imaginable that the formation of magnesite in the low-temperature deposits just mentioned became possible only under the influence of high dissolved CO_3^{2-}, even though rather long times were available, in comparison with the duration of most experiments.

Returning to Fig. 32, we may state that the growth of magnesite is practically obstructed, at ordinary temperature and for times normally available in laboratory experiments, by the water dipoles attached to the surface cations. Using the vernacular of catalyst research, one may say that the surfaces of magnesite nuclei are "poisoned" by firmly adsorbed water molecules.

According to the kinetic interpretation of the law of mass action, a relation should exist between the solubility product k_s and the rate constants for crystallization k_{cryst} and dissolution k_{diss}:

$$(k_s)^2 = \frac{k_{diss}}{k_{cryst}}.$$

However, notwithstanding the general validity of the law of mass action, which is safely based on thermodynamics, the kinetic interpretation is valid only when dissolution and crystallization proceed according to analogous mechanisms, i.e. when the rate of crystallization is proportional to the product of the concentrations $[Mg^{2+}] \cdot [CO_3^{2-}]$ and when, at the same time, the rate of dissolution is inversely proportional to the same product. The picture proposed in Fig. 32 for the surface conditions of a magnesite crystal in solution is indeed compatible with the latter type of dependence of the rate of dissolution: high concentrations of either CO_3^{2-} or Mg^{2+} would both suppress dissolution. However, according to the same picture, the rate of crystallization is more strongly favored by $[CO_3^{2-}]$ than by $[Mg^{2+}]$. It is even likely that, above a certain level of $[Mg^{2+}]$, $[CO_3^{2-}]$ is the only parameter determining the growth rate.

Most experimental determinations of the solubility of magnesite have been attempted in solutions saturated with CO_2, i.e. at very low $[CO_3^{2-}]$, and have yielded discordant values for the solubility product (HALLA and VAN TASSEL, 1964) or have failed to reach equilibrium (GARRELS, THOMPSON and SIEVER, 1960). This behavior must be expected when dissolution and crystallization are characterized by different mechanisms. If, in the course of dissolution, the solubility for a proposed set of parameters (temperature, CO_2 pressure etc., see below on pp. 103 and 162) is exceeded because of fluctuations of these parameters, the material dissolved in excess has no chance to become redeposited on the crystalline phase, because the rate of this process is much too slow, in comparison with the rate of dissolution, at low $[CO_3^{2-}]$.

Solubility normally refers to undeformed crystals of practically infinite grain size. Natural samples as used in the experiments of the authors just mentioned had to be ground, and thus contained excessively small grains and deformed crystals, or at least crystals containing distorted domains. Such material must be expected to dissolve in excess over the normal solubility. This influence has been observed and discussed by GARRELS et al. (1960), who were among the first to ascribe the failure in obtaining equilibrium for the solubility of dolomite to such factors as make for irreversibility also in the dissolution of magnesite. It is remarkable that a definite experimental solubility for magnesite has been determined by HORN (1969). This author used synthetic magnesite of such grain size that intense grinding was not required (private communication). His experimental solubility product $k_{sM} = 10^{-8.14 \pm 0.06}$ for 25° C is in excellent agreement with the previously quoted value of $10^{-8.1}$, which LANGMUIR derived from the thermodynamic properties of magnesite.

In summary, under earth-surface conditions, the precipitation of magnesium as a carbonate is characterized by the metastable formation of hydromagnesite (or nesquehonite). These hydrous carbonates form, since the last water ligands of the magnesium, whose removal would require activation energy, are incorporated into the crystal structures. They may convert very slowly, if at all, to anhydrous magnesite, and there is evidence, both observational (ALDERMAN; VON DER BORCH) and theoretical (expounded above), that this conversion may be accelerated by carbonate ions. The rôle of the carbonate in solution is twofold: not only must the solubility product $[Mg^{2+}] \cdot [CO_3^{2-}] = k_{sM}$ be exceeded, but it has to be exceeded in such a way that $[CO_3^{2-}]$ prevails over $[Mg^{2+}]$. For, only in the presence of as many activated CO_3^{2-} ions as possible, is it conceivable that magnesite may grow at perceptible rates near earth-surface temperatures, according to the ideas developed above.

For sketching the stability relations in the system $MgO–CO_2–H_2O$, LANGMUIR's fig. 5 may be used as the starting point. However, the field of magnesite has to be enlarged to such an extent that magnesite is the only stable phase besides brucite. The boundary between these two phases is determined by $P_{CO_2} = 10^{-6.2}$ for 25° C, as deduced by LANGMUIR from the thermodynamic solubility product $k_{sM} = 10^{-8.1}$. For short reaction times, hydromagnesite, nesquehonite, lansfordite, and brucite then form metastable equilibria within the stability field of magnesite. Artinite may have a small metastable field, perhaps wedging out toward high temperature, along the metastable boundary brucite-hydromagnesite. The revised phase diagram for the magnesium carbonates is shown in Fig. 33.

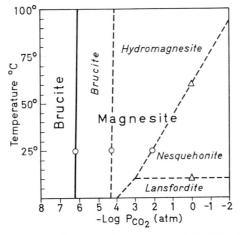

Fig. 33. Tentative phase diagram of the system $MgO–CO_2–H_2O$, modified from LANGMUIR (1965). Metastable fields are delimited by dashed lines and marked by the names of the metastable phases in italics. Triangles refer to direct experimental data, circles to thermodynamic data

It is worth mentioning that magnesite is now the only simple anhydrous magnesium salt besides MgF_2, sellaite, which is thermodynamically stable at earth-surface temperatures. However, in contrast to magnesite, sellaite forms instantaneously from aqueous solutions at ordinary temperature, even though as a very fine-grained precipitate. The ease of formation of anhydrous sellaite may be explained by the fact that the F^- ion is the smallest existing anion, being about equal in size to the water dipole. It is thus understandable that the F^- ion can more easily supplant any water-dipoles bonded to magnesium than do larger anions such as CO_3^{2-}.

c) Hydrated Calcium Carbonates

Hydrous carbonates of iron and manganese appear to be non-existent, as has been mentioned above, although they might be expected on account of the small size of these cations and the ensuing high interaction energies with the water molecule (Tables 20 and 21). In contrast, two well defined hydrated carbonates of calcium, $CaCO_3 \cdot 6H_2O$ and $CaCO_3 \cdot H_2O$, have been found to form in experiments aimed at the precipitation of calcium carbonate from aqueous solutions, e.g. by BROOKS, CLARK and THURSTON (1950). Hydrates intermediate between these two compounds have been claimed to occur as minerals, and their occurrences have been reviewed in PALACHE, BERMAN and FRONDEL. They have not been confirmed so far by modern diagnostic methods, and it is possible that they are in reality partially dehydrated $CaCO_3 \cdot 6H_2O$.

This latter compound is obtained in the laboratory preferably at temperatures near the freezing point of water (JOHNSTON et al., 1916), but it may form also near room temperature in solutions containing "calgon" (sodium polyphosphate) according to BROOKS et al. These authors explain the action of calgon (and of other additives) in bringing about the crystallization of $CaCO_3 \cdot 6H_2O$ by the tendency of the polyphosphate to be adsorbed on the surface of the nuclei of calcite whose growth is thus inhibited. At any rate, $CaCO_3 \cdot 6H_2O$ and also $CaCO_3 \cdot H_2O$ tend to convert to anhydrous carbonates, such as calcite, even at low temperatures, when the additives which caused their formation are no longer present. Hydrated calcium carbonates thus appear to have no thermodynamic stability field, but they form for kinetic reasons. This situation is completely analogous to the metastability, relative to magnesite, of hydromagnesite and nesquehonite, although the rate of conversion to the stable anhydrous carbonate is considerably greater for calcium than for magnesium, in line with the trend in the cation-water interaction energies.

The additive provoking the crystallization of $CaCO_3 \cdot H_2O$ is the magnesium ion. Its rôle is perhaps the same as in the formation of aragonite, to be discussed later (pp. 107—116), in that the growth of the more stable calcite is inhibited in the presence of sizable amounts of dissolved magnesium.

The metastable formation of the hydrated carbonate must be related to the fact that the last water molecule, whose removal should be the most difficult, need not be separated from the cation, and the corresponding activation barrier need not be overcome. Temperature does not appear to be as critical as in the case of $CaCO_3 \cdot 6H_2O$, since $CaCO_3 \cdot H_2O$ has been obtained up to 40° C by BROOKS et al. At 20° C a molar ratio Mg/Ca of about 4.5 or higher is required for $CaCO_3 \cdot H_2O$ to form (LIPPMANN, 1959; VAN TASSEL, 1962). The Mg/Ca ratio in sea water is about 5, and $CaCO_3 \cdot H_2O$ has been precipitated from sea water by the addition of Na_2CO_3 solution (MONAGHAN and LYTLE, 1956; BARON and PESNEAU, 1956). According to this experience, $CaCO_3 \cdot H_2O$ may be expected to form from sea water under certain natural conditions also, and it may have so far escaped detection. In view of this prospect microphotographs of characteristic artificial crystals are presented in Figs. 34 and 35, as an aid in the identification of possible natural crystals formed from sea water. It must be borne in mind, however, that in most instances artificial $CaCO_3 \cdot H_2O$ has been obtained in the form of spherulites, except in the experiments of the present author (LIPPMANN, 1959), from which the crystals shown in Figs. 34 and 35 originate.

The first reported natural occurrence of $CaCO_3 \cdot H_2O$ is in Lake Issyk-Kul in the Soviet Union where it forms incrustations on the bottom near the shore (SAPOZHNIKOV and TSVETKOV, 1959; E. I. SEMENOV, 1964). The Mg/Ca ratio in the water of this lake is 3.5, and the high content of bicarbonate noted is not sufficient to precipitate all of the dissolved calcium as a carbonate (SAPOZHNIKOV and VISELKINA, 1960). The temperature at a depth of 5 meters is $18.3 \pm 1.3°$ C according to KEISER

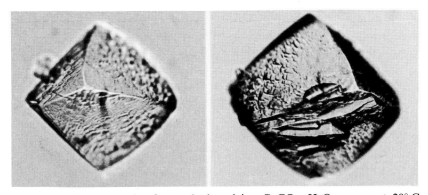

Fig. 34. Artificial crystals of monohydrocalcite, $CaCO_3 \cdot H_2O$, grown at 20° C from a solution containing 60 mg $CaSO_4 \cdot 2H_2O$, 1000 mg $MgSO_4 \cdot 7H_2O$ and 2000 mg KOCN per liter. From the latter compound carbonate is developed by hydrolysis (LIPPMANN, 1959). The two crystals shown are enantiomorphous (left- and right-handed) trapezohedra according to the symmetry $D_3 - 32$. The refractive indices are $n_\varepsilon = 1.543$, $n_\omega = 1.590$. The birefringence is thus lower than in calcite and vaterite. From this, it must be concluded that the planes of the CO_3 groups in $CaCO_3 \cdot H_2O$ are neither parallel to the c axis as in vaterite, nor perpendicular to it as in calcite. The possible space groups determined for the single crystals are $D_3^{(4,6)} - P3_{(1,2)}21$ or $D_3^{(3,5)} - P3_{(1,2)}12$; they are compatible with a tilted arrangement of the CO_3 groups. Scale: the horizontal dimension of the left crystal is 0.2 mm

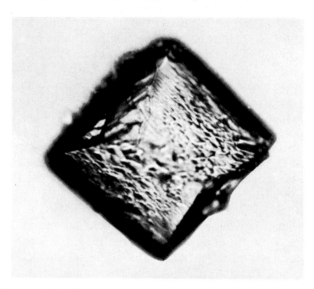

Fig. 35. View on the short equatorial edge of a right-handed trapezohedral crystal of $CaCO_3 \cdot H_2O$. During the early stages of the study (LIPPMANN, 1959), the short edges were attributed to distorted crystal growth when they were found to occur on some crystals but missing on others, and the crystals were inappropriately described as trigonal bipyramids. However, the arrangement of the vicinals unequivocally hinted at trapezohedral symmetry. The examination of more crystals has revealed that the short equatorial edges occur more often than not and that most of the crystals are thus indeed trigonal trapezohedra. Scale: the horizontal dimension of the crystal is 0.2 mm

(1928). E. I. SEMENOV, to whom we owe the definite identification of $CaCO_3 \cdot H_2O$ in Lake Issyk-Kul, has proposed the mineral name mono-hydrocalcite in allusion to the composition. CARLSTRÖM (1963) found microscopic trigonal bipyramids of $CaCO_3 \cdot H_2O$ among the statoconia in the labyrinth of the tiger-shark, *Galeocerdo*, in addition to aragonite and calcite. FISCHBECK and MÜLLER (1971) report the mineral from speleothems of a cave in Jurassic dolomites in Franconia, Germany.

The formation of hydrated calcium carbonate at earth-surface temperature shows that the dehydration of the cation may still be an obstacle for the crystallization of anhydrous calcium carbonate. This is perhaps the reason why barium is usually preferred over calcium, when carbonate is to be precipitated in the laboratory for the purpose of CO_2 determinations. A hydrated carbonate of bariums is not known to exist, and it may be conceived that dehydration is no longer rate-determining in the growth of anhydrous salts of a cation as large as barium. It is possible that the 23.9 kcal/mol, required for the removal of the last water dipole from Ba^{2+}, are about balanced by the energy gained by the adsorption of CO_3^{2-} on the surface of a carbonate crystal. Barium would then mark the zero in a scale of hydration-conditioned activation energies pertaining to cations adsorbed on carbonate crystal surfaces. Cation-water interaction energies for the bivalent cations, diminished by the energy for barium, are given in Table 21, column 2. As long as data from more immediate sources are not available, these values may serve to give an idea of the minimum energies to be expected for the dehydration barriers inhibiting the growth of anhydrous carbonate crystals. The values given in column 1 would be representative of the maximum energies. Table 21 thus shows the possible range of activation energies determining the growth rates of anhydrous carbonates.

The existence of two basic carbonates for a cation as large as lead, $2PbCO_3 \cdot Pb(OH)_2$ (hydrocerussite) and $6PbCO_3 \cdot 3Pb(OH)_2 \cdot PbO$ (plumbonacrite; OLBY, 1966), is hard to explain from the point of view of purely ionic bonding and is perhaps related to the tendency of lead to be bonded via non-polar forces.

3. Alkali-Bearing Carbonates

In addition to the carbonate minerals treated in the preceding chapters, all of which are characterized by bivalent cations, there occur also in the sedimentary cycle a number of carbonates of monovalent cations, i.e. alkali carbonates. They form preferably by evaporation from continental waters of appropriate composition, notably in arid regions. Some of these carbonates, such as trona, thermonatrite, soda, and gaylussite have been known as minerals from the beginnings of mineralogy (PALACHE et al.). All of the alkali-bearing carbonate minerals which have been described so far are listed in Table 22. They have become known chiefly from the recent sediments of saline lakes, e.g. in Egypt, Venezuela, and in the western part of the United States. Although such occurrences have been discussed in treatises of geochemistry, as e.g. by CLARKE (1924), their geochemical importance does not seem to have been duly appreciated, perhaps because it is difficult to estimate the overall abundance of the alkali carbonates. Lately, important ancient deposits of these minerals have been located in the subsurface in the states of Wyoming, Utah and Colorado. They occur as thick beds intercalated in the dolomitic sediments of the lacustrine Green River formation, which is of Eocene age. Trona is mined from them in Wyoming and contributes a sizable portion to the soda ash production of the United States (FAHEY, 1962). Thus at least in the Green River formation, the alkali-bearing carbonates designated by 1; G in Table 22 are of rock-forming importance. In view of their massive occurrence at least in this one region, it may be expected that such minerals have been present, although in smaller amounts, in other sediments of lacustrine origin. They may have disappeared, either during diagenesis by reaction with solutions contain-

Table 22. Alkali-carbonate minerals

Chemical formula	Name	Reference
Acid carbonates (bicarbonates)		
$NaHCO_3$	nahcolite	1; G
$KHCO_3$	kalicinite	
$Na_2CO_3 \cdot NaHCO_3 \cdot 2H_2O$	trona	1; G
$Na_2CO_3 \cdot 3NaHCO_3$	wegscheiderite	2; G
Anhydrous normal carbonates		
$Na_2Ca_2(CO_3)_3$	shortite	1; G
$Na_2Mg(CO_3)_2$	eitelite	G
$Na_2(Ca, Ba, Sr, La, Ce)_4(CO_3)_5$	burbankite	G
Hydrated normal carbonates		
$Na_2CO_3 \cdot H_2O$	thermonatrite	3; G
$(Na_2CO_3 \cdot 7H_2O)$	(unknown as a mineral?)	
$Na_2CO_3 \cdot 10H_2O$	soda	G
$Na_2Ca(CO_3)_2 \cdot 2H_2O$	pirssonite	1; G
$Na_2Ca(CO_3)_2 \cdot 5H_2O$	gaylussite	1; G
Carbonates containing hydroxyl or halogen		
$NaAl(CO_3)(OH)_2$	dawsonite	4; G
$Na_3Mg(CO_3)_2Cl$	northupite	1; G
Compound carbonates		
$Na_6Mg_2(CO_3)_4(SO_4)$	tychite	1; 5; G
$Na_3Mg(CO_3)(PO_4)$	bradleyite	1; G
$Na_6(CO_3)(SO_4)_2$	burkeite	
$Na_{22}K(SO_4)_9(CO_3)_2Cl$	hanksite	

Explanation of references:
1 FAHEY and MROSE (1962)
2 FAHEY and YORKS (1963); APPLEMAN (1963)
3 SWANSON, GILFRICH, COOK, STINCHFIELD and PARKS (1959)
4 SMITH and MILTON (1966)
5 KEESTER, JOHNSON and VAND (1969)
G means that the mineral occurs in the Green River formation and that the papers of MILTON and FAHEY (1960) and MILTON and EUGSTER (1959) contain or lead to more information. All of the alkali carbonate minerals, as far as they were known up to about 1950, i.e. with the exception of wegscheiderite, eitelite and burbankite, are also reviewed in PALACHE, BERMAN and FRONDEL (1951). More references are, of course, available, but those given here will lead to the original description and, where available, to reliable X-ray diffraction powder data. For the structure of eitelite see: PABST, A.: Am. Mineralogist **58**, March-April 1973.

ing calcium and/or magnesium to yield more insoluble carbonates, such as calcite or dolomite, or have been removed by leaching, because of their rather high solubilities.

The carbonates containing calcium in addition to sodium dissolve differentially leaving a residue of calcite, sometimes as pseudomorphs after the original mineral. Thus pseudomorphs after shortite occur in the outcrops of the Green River formation (FAHEY, 1962). Certain calcareous pseudomorphs occurring in recent marine or brackish muds have been attributed to gaylussite and termed pseudogaylussite (see PALACHE et al., pp. 160—161). The formation of pseudomorphs is a rare happening, requiring euhedral crystals to begin with and then special conditions of transformation favoring the preservation of the original crystal shape (see CORRENS, p. 309). The contribution of sodium calcium carbonates to ancient sediments, especially to those of lacustrine origin, may thus be greater than is now realized.

To trace the origin of alkali carbonates we must consider weathering. The alkali is leached from appropriate silicates and becomes balanced in solution by OH^- ions through hydrolysis, which occurs as a result of the rather low solubility of the silicate anion. The OH^- reacts with the carbon dioxide of the air according to

$$OH^- + CO_2 \rightarrow HCO_3^- \tag{7}$$

or

$$2\,OH^- + CO_2 \rightarrow CO_3^{2-} + H_2O \tag{8}$$

to yield bicarbonate and/or carbonate depending on concentrations. The resulting alkali carbonate solutions may become concentrated by evaporation under conditions of poor or no drainage, giving rise to features ranging from alkali crusts to alkaline lakes. In some instances, (hot) springs supply alkali carbonate. This may be derived by leaching from the alkali carbonate content of buried sediments. In the case of juvenile waters, the ultimate source, as in weathering, is from silicates from which alkali may become mobilized by metamorphic processes. The fact that the main cation of the sedimentary alkali carbonates is sodium and that the only potassium carbonate mineral, kalicinite, is extremely rare, is most probably related to the preferential fixation of potassium in clay minerals, notably illite.

As for the crystal structures of alkali-bearing carbonates, it would be beyond the scope of the present monograph to discuss in detail even the incomplete information thus far available. Nahcolite (SASS and SCHEUERMANN, 1962) and kalcinite (see PEDERSEN, 1968) seem to be the only alkali carbonate minerals for which modern structure refinements are available. In addition, the structures of thermonatrite (HARPER, 1936)

and trona (BROWN, PEISER and TURNER-JONES, 1949) have been analyzed. Among the structures of complex alkali-bearing carbonates, those
of dawsonite (FRUEH and GOLIGHTLY, 1967), pirssonite (CORAZZA and
SABELLI, 1967), and gaylussite (MENCHETTI, 1968) are well known. In the
latter two minerals, the cations are characterized by analogous coordination numbers with respect to oxygen: NaO_6 and CaO_8. In pirssonite
we have: $(H_2O)_2NaO_4$ and $(H_2O)_2CaO_6$; in gaylussite: $(H_2O)_2NaO_4$ and
$(H_2O)_4CaO_4$. Thus, as in other hydrated crystals, water molecules are
directly bonded to the cations. The hydration of the calcium in these two
minerals may be the reason why it has not been possible so far to
synthesize the anhydrous mineral shortite at room temperature.

No serious kinetic obstacles are experienced in the system Na_2CO_3–
$NaHCO_3$–H_2O (Fig. 36), perhaps because most of the stable phases involved near earth-surface conditions are hydrated, so that in their crystallization the dehydration barrier need not be overcome. Moreover, the
interaction energy of monovalent sodium with the water molecule is

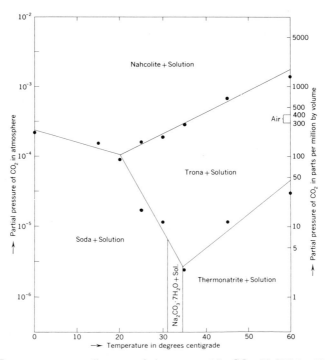

Fig. 36. P_{CO_2}-temperature diagram of the system Na_2CO_3–$NaHCO_3$–H_2O after
MILTON and EUGSTER from FAHEY (1962). Wegscheiderite, $Na_2CO_3 \cdot 3\,NaHCO_3$,
has been prepared at a minimum temperature of 89.5° C (WEGSCHEIDER and MEHL,
1928), i.e. outside this diagram, where it separates the fields of nahcolite and trona

about half that of bivalent calcium (Table 20, column 3). This energy barrier is obviously not too critical, judging from the relative ease with which anhydrous nahcolite crystallizes in its stability field. In contrast, the dehydration barrier of the calcium may have affected the phase diagram Na_2CO_3–$CaCO_3$–H_2O (Fig. 37), in which only hydrated phases occur, whereas anhydrous shortite is missing. The occurrence in the Green River formation does not supply us with any evidence that shortite would require any more extreme conditions of formation than the other alkali-bearing carbonates associated with it. Nevertheless, anhydrous shortite has so far eluded artificial preparation below 125° C (BRADLEY and EUGSTER, 1969). This may be due to an unusually slow rate of crystallization, caused by the dehydration barrier of the calcium. It is thus possible that the hydrated phases, gaylussite and pirssonite, appear metastably instead of anhydrous shortite in at least part of their formation fields in Fig. 37. Such a situation would be analogous to the metastable formation of hydrous magnesium carbonates instead of magnesite.

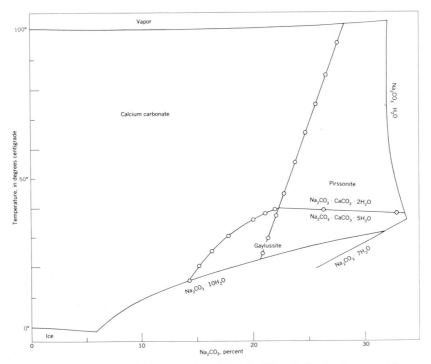

Fig. 37. Phase diagram of the system Na_2CO_3–$CaCO_3$–H_2O after BURY and REDD from FAHEY (1962). Shortite, $Na_2CO_3 \cdot 2CaCO_3$ may be a stable phase in part of this diagram, but has not so far been synthesized below 125° C

The problem may perhaps be settled by stability data derived from the thermodynamic properties of the sodium calcium carbonates. Reliable information concerning the structure of shortite might help to corroborate the above proposed kinetic explanation of the failure to synthesize the mineral at lower temperatures.

The chemical formulae of the alkali-bearing carbonates of the sedimentary cycle have been presented in Table 22 chiefly for the purpose of explaining the mineral names used in the text. However, the chemical formulae also suggest the following: These minerals occur in many areas characterized by poor drainage and they vary in abundance from deposits of economic importance, such as in the Green River formation, to inconspicuous efflorescences. For all these occurrences, saturated solutions are required. Given the rather high solubilities of the alkali carbonates, it may be concluded that undersaturated solutions, though still containing appreciable amounts of alkali carbonates, are more widespread than is realized normally. Such solutions must be regarded as potentially important in the formation and transformation of calcareous sediments on the continents. When discharged into restricted bodies of marine water, such as lagoons, solutions of alkali carbonate will promote the precipitation of calcium carbonate and they may be instrumental in the diagenesis of deposited carbonate. The first example for the interaction of continent-derived alkali-carbonate waters with sea water has been pointed out by VON DER BORCH (1965) and ALDERMAN (1965) who ascribe the formation of recent dolomite in the lagoons of South Australia to this influence. Similar settings may occur along the coasts of other arid regions.

A general picture of the abundance of alkali carbonates is obtained when the alkali and alkaline-earth contents of river and lake waters are subdivided as to how they are balanced, by strong anions, that is, by chloride and sulphate, or by carbonate and bicarbonate. The last pair may be lumped together as "alkalinity". Based on the compilation of LIVINGSTONE (1963), MACKENZIE and GARRELS (1966) came to the conclusion that about 75% of the bicarbonate discharged by the world's rivers into the sea is balanced by calcium and will be precipitated ultimately as calcium carbonate. The remaining bicarbonate is balanced chiefly by alkalis and magnesium. Thus considerable alkalinity is set free on the continents, chiefly by weathering. When discharged into poorly drained continental basins, rather than the sea, the alkalinity in excess over calicum may enter into various reactions and, under favorable conditions, give rise to deposits of alkali-bearing carbonates. More work is desirable in the field of the "Geochemistry of Alkalinity" in continental environments.

C. The Polymorphism Calcite-Aragonite

I. Stable Relationships

The occurrence of calcium carbonate as two different minerals, calcite and aragonite, was one of the earliest and most discussed examples of polymorphism. The early publications concerning the problem have been reviewed by GOSSNER (1930) in HINTZE's Handbuch der Mineralogie. However, the early history of the problem has not by any means developed in a straightforward fashion and has lost much of its interest when seen in the light of the present state of knowledge.

LE CHATELIER (1893) was the first to use a modern physicochemical approach to the polymorphism calcite-aragonite, by considering the differences in density and heat content. When heated, dry aragonite converts to calcite irreversibly around 400° C in an endothermic reaction. This is nowadays most conveniently observed by the technique of differential thermal analysis (DE KEYSER and DEGUELDRE, 1950; FAUST, 1950). Using the rule which now bears his name, LE CHATELIER concluded from the higher heat content of calcite relative to aragonite, that aragonite might be the stable polymorph of $CaCO_3$ at low temperatures. However, the stability range must be far below the conversion temperature of about 300°—400° C because of the irreversibility of the reaction. From the higher density of aragonite, LE CHATELIER also concluded that aragonite should be stabilized by pressure.

From quantitative thermodynamic data, BÄCKSTRÖM (1925) calculated that calcite should convert to aragonite at 25° C under a pressure of more than 2900 atmospheres ($= 2.94$ kilobars), and that at one atmosphere aragonite might become stable at very low temperatures near absolute zero. Later, BÄCKSTRÖM's results were essentially confirmed by thermodynamic calculations carried out by JAMIESON (1953) and by MACDONALD (1956) on the basis of new data. These authors were also the first to verify experimentally the status of aragonite as a high-pressure phase. JAMIESON, by means of the electrolytic conductivity, measured the solubility of calcite and aragonite in water, as a function of pressure at a number of temperatures below 100° C. At those pressures he identified the equilibrium at which both minerals imparted the same electrolytic conductivities to their solutions. JAMIESON's results are included in Fig. 38. MACDONALD followed the conversion directly for temperatures between 300° and 600° C in a "simple squeezer", and the

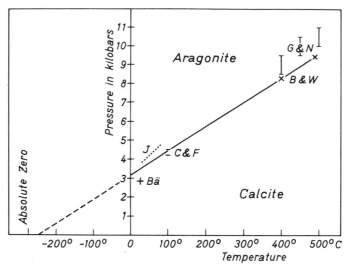

Fig. 38. The phase diagram calcite-aragonite summarizing results of various authors. *Bä:* Thermodynamic calculation of BÄCKSTRÖM (1925). *J* (dotted line): JAMIESON (1953); *C & F:* CRAWFORD and FYFE (1964); *G & N* (vertical bars): GOLDSMITH and NEWTON (1969); *B & W* (crosses): BOETTCHER and WYLLIE (1968). Data above 500° C, although available, are not shown because of the complications arising from the appearance of calcite II, which is of no consequence for the sedimentary occurrences of $CaCO_3$. For details concerning calcite II, see the last two references and JAMIESON (1957)

resulting phases were obtained for direct examination by suitable quenching. More recent direct experimental determinations of the equilibrium calcite-aragonite, especially those carried out by means of hydrothermal techniques (CRAWFORD and FYFE, 1964; BOETTCHER and WYLLIE, 1968; GOLDSMITH and NEWTON, 1968), have yielded equilibrium pressures which differ somewhat from the original results of JAMIESON and MAC-DONALD, but the general nature of the phase diagram established by these authors remains valid. A diagram, based on the latest data, and including some of the earlier determinations, is shown in Fig. 38.

When the phase boundary is extended as a straight line (dashed in Fig. 38) to low temperature it ends up near absolute zero. This means that the crystal energies of calcite and aragonite should be nearly identical and that the difference in energy at higher temperatures must be due chiefly to different mechanisms of thermal vibration. The higher heat content of calcite may be explained by the greater freedom of the CO_3 groups to vibrate, in contrast to aragonite, where the CO_3 groups are more densely packed and more constrained (see p. 59). More work, both experimental and theoretical, is needed to elucidate these points.

It is evident from the phase diagram, Fig. 38, that the stable formation of aragonite at the temperatures of the earth's crust is possible only under considerable pressures, of thousands of bars. Such conditions may occur during the formation of certain metamorphic rocks. Thus the aragonite occurring in the glaucophane schists and metasediments of the Franciscan formation of California appears to have formed under equilibrium conditions, since other minerals indicative of high pressure, such as glaucophane, jadeite, clinozoisite, and garnet, are present (COLEMAN and LEE, 1962; McKEE, 1962).

According to the phase diagram aragonite-calcite, and judging from the thermodynamic properties of these phases, calcite is the stable modification of calcium carbonate under earth-surface conditions. All of the aragonite forming under sedimentary conditions and in organisms must be regarded as thermodynamically metastable relative to calcite, to which it should ultimately convert. In particular, the metastability of aragonite is reflected by its solubility in water, which is higher than that of calcite: the solubility product

for aragonite is: $k_{sA} = 10^{-8.22}$

and for calcite : $k_{sC} = 10^{-8.35}$.

The values are those determined for 25° C and one atmosphere by GARRELS, THOMPSON and SIEVER (1960, 1961).

At this point, it must be mentioned that calcite converts to aragonite, at least partly, on prolonged grinding in mechanical mortars (BURNS and BREDIG, 1956; DACHILLE and ROY, 1960; JAMIESON and GOLDSMITH, 1960). This has been ascribed to the high pressures which are exerted when moderate forces act upon exceedingly small areas, as e.g. on the comminuted calcite crystallites. However, an alternative explanation is possible. By the grinding, the structure of the calcite crystallites may be disturbed to such a degree that the ground calcite becomes more soluble than aragonite in water. In this way, aragonite may form by precipitation in adhering moisture films (JAMIESON and GOLDSMITH, 1960; CHAVE and SCHMALZ, 1966).

Calcite and Aragonite-Type Structures in Systems Other than CaCO₃

The carbonates of the larger bivalent cations crystallize with the aragonite structure as explained on p. 64. In contrast to aragonite proper, the aragonite isotypes witherite, BaCO₃, and strontianite, SrCO₃, are stable under earth-surface conditions, where no other modifications are known. However, under normal pressure, BaCO₃ and SrCO₃ convert to calcite-

type polymorphs on heating above 800° and 910° C, respectively (LAN-
DER, 1949). These high-temperature polymorphs are not entirely identical
with the calcite structure. They show a c_{hex} period which is about half
that of calcite proper, as evidenced by the absence of $(hk.l)$ X-ray reflections with l odd, when the hexagonal cell dimensions similar to those of
calcite are chosen. These systematic extinctions have been explained by
LANDER in terms of rotational disorder of the CO_3 groups with respect
to the c_{hex} axis. As in the system $CaCO_3$, the high-temperature calcite-
type polymorphs of $BaCO_3$ and $SrCO_3$ have lower densities than the
low-temperature aragonite-type polymorphs. Thus the conversion must
be expected to shift to higher temperatures under elevated pressures in a
way analogous to the $CaCO_3$ system, although the conversions take
place at much higher temperatures than in the latter. $BaCO_3$ and $SrCO_3$
may, therefore, be regarded as models of $CaCO_3$ as far as the conversion
aragonite–calcite is concerned.

In the early history of the aragonite-calcite problem it was assumed
that the formation of aragonite might be engendered by strontium, since
some of the early chemical analyses of aragonites showed a few percent
of strontium, whereas calcites were found to be nearly free from Sr. Although ROSE (1837) showed that aragonite may be prepared in the laboratory in the absence of additions of strontium, the discussion that the
presence of strontium may cause the formation of aragonite in nature
was resumed from time to time.

LANDER found that the conversion temperature of $SrCO_3$ to the
calcite-type structure is lowered by the admixture of $CaCO_3$. As a crude
approximation, it may be assumed that the conversion temperature in
the system $SrCO_3$–$CaCO_3$ changes in an essentially linear fashion, between 910° C for pure $SrCO_3$ and the conversion temperature for
$CaCO_3$ near absolute zero (Fig. 38). Accordingly, about 25 mol% $SrCO_3$
in solid solution would be required to stabilize an aragonite-type mixed-
crystal at 25° C and one atmosphere. MACDONALD obtained 30 mol%,
and FYFE and BISCHOFF (1965) 15 mol%, from more conventional thermodynamic calculations. TERADA (1953) was able to prepare, at room
temperature, mixed crystals of calcite structure containing up to
10 mol% $SrCO_3$. Natural aragonites always contain less strontium, in
most cases below 1%, so it is inconceivable that such amounts may
stabilize the aragonite structure in minerals consisting predominantly of
$CaCO_3$. Therefore, even when the incorporation of strontium in natural
aragonites is considered, nearly all of them must be metastable formations, the only exception being those occurrences in metamorphic rocks
of high-pressure origin.

Before discussing possibilities for the metastable formation of aragonite under the conditions of the sedimentary cycle, it is interesting to

show how aragonite-type phases exist in competition with calcite-type phases in non-carbonate AXO₃ compounds.

Of the alkali nitrates, $LiNO_3$ and $NaNO_3$ are isotypes of calcite, whereas KNO_3 crystallizes with the aragonite structure at room-temperature. Thus the structure type appears to be determined by the size of the cation (see BRAGG and CLARINGBULL, p. 131; CORRENS, p. 62), just as in the alkaline-earth carbonates. Analogous to the latter, KNO_3 has a high-temperature polymorph of calcite structure which forms above 128° C. On passing this temperature, both ways, the conversion may be conveniently studied under the microscope (CORRENS, p. 187—188), and such observations may serve to illustrate the phase transformations in the system calcite–aragonite.

The rare-earth borates, together with the borates of other trivalent cations, represent another group of ABO_3 compounds which contain models for the structures of calcite and aragonite (GOLDSCHMIDT and HAUPTMANN, 1931). Whereas, among the carbonates, the vaterite phase is unique to $CaCO_3$ and is unstable relative to both calcite and aragonite, the stable crystallization of vaterite-like structures is common among the rare-earth borates. Aragonite and calcite structures are restricted to the borates of the largest (La^{3+}, Nd^{3+}) and the smallest cations (Lu^{3+}, In^{3+}), respectively, whereas the intermediate range of cation sizes is characterized by the vaterite structure. These results were obtained by LEVIN, ROTH and MARTIN (1961) [see also ROTH, WARING and LEVIN (1964)] and are illustrated in Fig. 39. BRADLEY, GRAF and ROTH (1966) have proposed plausible structures for the vaterite-type rare-earth borates, notably for $YbBO_3$, from which conclusions concerning the possible structure for vaterite proper ($CaCO_3$) have been drawn above (Fig. 29 and 30). For a comparison of the dependence of structural stability upon cation size, account must be taken of the different X–O distances in the BO_3 anion (1.35 Å, BRADLEY, GRAF and ROTH) and in the CO_3 group (1.28 Å). Judging then from the rare-earth borates, $CaCO_3$ should show the aragonite and the vaterite structures. The calcite structure should occur only for a carbonate of a cation at least as small as Mn^{2+}, which would, at the same time, mark the limit of stability of the vaterite structure. However, the stability of calcite proper, the metastability of vaterite proper, and the non-existence of vaterite-type $MnCO_3$ show that size relations are reliable only as qualitative indicators of existing polymorphs and that their stability is influenced considerably by such differences in charge distributions as are expressed by the formulae $Ca^{2+}(CO_3)^{2-}$ and $Yb^{3+}(BO_3)^{3-}$. The situation is complicated by the fact that the vaterite structure among the rare-earth borates undergoes high-low inversions on heating and cooling (Fig. 39). In the low-temperature form, the BO_3 unit shows a strong pyramidal deformation, in contrast to

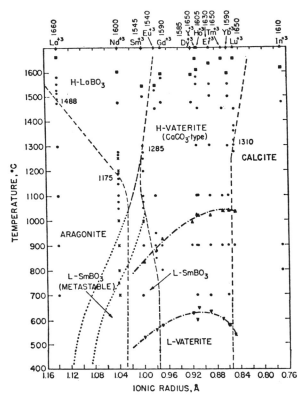

Fig. 39. Stability relationships and occurrence of the structures of aragonite, vaterite and calcite among the borates ABO_3 of the rare earths and related trivalent cations as a function of temperature and the ionic radius (AHRENS) of A^{3+} (from ROTH, WARING and LEVIN, 1964). ● subsolidus experiments, ■ liquidus temperatures (given along the top of the figure), × high-temperature X-ray diffraction experiments, ▲ inversions by D.T.A. on heating of vaterite polymorph, ▼ inversions by D.T.A. on cooling of vaterite polymorph. Note the occurrence of additional structures typified by high-temperature (H-) $LaBO_3$ and low-temperature (L-) $SmBO_3$

the high-temperature form where it is nearly planar (BRADLEY, GRAF and ROTH). The pyramidal deformation of BO_3 may be attributed to the greater charge of the trivalent cations compared to bivalent calcium. Nevertheless, the comparison of borates and carbonates is extremely instructive, especially in view of other existing analogies: Dolomite-type double borates appear to be represented by the rare mineral nordenskiöldine, $Ca^{2+} Sn^{4+} (BO_3)_2$, its artificial Mg analogue (ALÉONARD and VICAT, 1966), and possibly by $LaLu(BO_3)_2$ (ROTH et al.). A number of huntite-type complex borates of the general formula $XAl_3(BO_3)_4$ have been synthesized also, where X is a rare earth (MILLS, 1962; BALLMAN, 1962). Their crystallographic data are in harmony with the symmetry postulated by GRAF and BRADLEY for huntite, $CaMg_3(CO_3)_4$. Thus the structural interpretation of the mineral by these authors is confirmed by independent information from isomorphous compounds.

II. The Metastable Occurrence of Aragonite in Aqueous Solutions at Normal Pressure

1. In the Absence of Bivalent Cations Other than Ca^{2+}.

There are two general ways for precipitating calcium carbonate from aqueous solutions. The first and simplest method consists of mixing solutions, one of which contains a dissolved calcium salt (chloride, nitrate, or sulphate), and the other a soluble carbonate, such as an alkali or ammonium carbonate. In the second method, calcium carbonate is first dissolved in water saturated with carbon dioxide. From such a solution, the calcium carbonate is reprecipitated when the solution is allowed to lose its carbon dioxide to the surrounding air. This second mechanism of calcium carbonate precipitation simulates natural conditions more closely than the first, since it is found to be active in most inorganic carbonate precipitations of the sedimentary cycle. It is said that calcium carbonate is dissolved as bicarbonate, although a solid bicarbonate of calcium is not known to exist. Actually the solubility of calcium carbonate, and of other sparingly soluble carbonates, under the influence of dissolved CO_2, is governed by the solubility product (see above, p. 99) and the interaction of the following equilibria involving the various species related to carbonic acid:

$$\frac{P_{CO_2}}{[H_2CO_3]} = 10^{+1.47} \quad \text{(P in atmospheres)}, \tag{9}$$

$$\frac{[H^+] \cdot [HCO_3^-]}{[H_2CO_3]} = 10^{-6.35}, \tag{10}$$

$$\frac{[H^+] \cdot [CO_3^{2-}]}{[HCO_3^-]} = 10^{-10.33}. \tag{11}$$

(The equilibrium constants to the right are for 25° C. Information concerning their variability with temperature is found in GARRELS and CHRIST, p. 89.)

The application of these equilibrium relations to the practical calculation of the solubility of calcium carbonate is discussed in GARRELS and

CHRIST, pp. 74—92, for a number of typical examples (see also p. 162 and v. ENGELHARDT, 1973).

Both methods for the laboratory preparation of calcium carbonates, as outlined above, were used by ROSE as early as 1837. This author found that calcite is obtained at room temperature, whereas aragonite forms preferably in hot solutions. The general rule that more and more aragonite forms instead of calcite as the solution temperature approaches 100° C has been confirmed in a large number of experiments by later authors, and the rule is quoted in most pertinent texts. However, limiting temperatures for the formation of aragonite and calcite in the range of 30° to 60° C are usually given with more authority than appears justifiable in the light of experience. Actually, such temperature limits are extremely sensitive to experimental detail, and conditions may be devised to spread their range considerably.

Progress since 1837 concerning the temperature dependence of the formation of calcite versus aragonite consisted chiefly in the increasing reliability of the determinative methods which became available for the identification of the precipitates. Whereas, during the time from ROSE (1837) to VATER (1893—1899), the main criteria were micromorphology and density, MEIGEN (1903, 1907, 1911) could use his staining test, which involves boiling in a 0.1 M solution of cobalt nitrate. JOHNSTON, MERWIN and WILLIAMSON (1916) relied chiefly on optical properties determined under the petrographic microscope.

With the availability of X-ray diffraction as a tool for quantitative phase analysis, it became possible to plot the percentages of the modifications obtained as a function of the precipitation temperature (DE KEYSER and DEGUELDRE, 1950; WRAY and DANIELS, 1957; KITANO, 1962a). The resulting diagrams are complicated by the occurrence of vaterite, which may form in addition to calcite and aragonite, usually above about 50° C, and they differ from each other because of differences in experimental conditions. Similar differences in the detailed results occurred also in the earlier publications. They are to be ascribed to different concentrations of the solutions used and to different modes and rates adopted in mixing the solutions, or in allowing escape the CO_2. The influence of these parameters is of about the same order as that of the presence or absence of neutral salts of monovalent cations in varying amounts, or of whether these are chlorides or sulphates. Sulphates have been claimed to promote the formation of aragonite. However, calcite may be obtained in the presence of sulphates even at 100° C, provided the precipitation is carried out at a sufficiently slow rate.

In general, the concentrations of the order 1 M to 0.1 M used in the method of mixing two solutions are too great to apply conclusions from such experiments to the natural formation of the $CaCO_3$ modifications.

However, the preferred formation of aragonite near the boiling point of water applies also to the method of CO_2 loss from calcium bicarbonate solutions, and this technique may be considered to simulate the natural conditions of aragonite deposition from thermal springs, especially when Ca concentrations of a realistic order (0.003 M) are used (KITANO, 1962a).

The tendency of aragonite to form preferably from hot solutions is at variance with the stability relations in the system $CaCO_3$ (Fig. 38). Aragonite is thermodynamically unstable even at ordinary temperature and becomes still more so with increasing temperature. Thus the formation of the less stable and more soluble aragonite can be explained only in terms of rates of nucleation and crystallization in comparison with the more stable calcite, whose solubility is exceeded even more than that of aragonite. The fact that the formation of aragonite may be more prevalent in solutions supersaturated with regard to both phases is indicative of a much higher crystallization rate for aragonite than for calcite. This relation of crystallization rates applies not only to hot solutions, for aragonite may form as the only phase also at room temperature, when seed crystals of aragonite are added under conditions which would otherwise yield calcite. On the other hand seeding with calcite is ineffective under conditions which would otherwise yield aragonite (LIPP-MANN, unpublished).

The greater crystallization rate of aragonite may be related to the greater number of oxygens (nine) surrounding the calcium in this structure, in constrast with six in calcite. The effect may be that incoming calcium ions are more easily captured by aragonite nuclei and, once adsorbed, are more effectively held back than by nuclei of calcite. The same would apply to incoming CO_3^{2-} groups when it is considered that their thermal vibrations are more restrained in aragonite than in calcite. This factor would entail a more efficient capture and retention of both CO_3^{2-} and Ca^{2+} by the active sites of aragonite nuclei. These speculations require support from calculations on the basis of the detailed crystal structures, and vaterite should be included in such considerations.

The increasing yield of aragonite with increasing temperature of precipitation implies, in addition to a greater crystallization rate, that the rate of spontaneous nucleation increases more sharply with temperature than for calcite. Thus a reversal of the relative importance of nucleation rates must occur between room temperature and $100°$ C. This phenomenon, too, should be susceptible to an interpretation in terms of the crystal structures. The only explanation thus far proposed is that by GOTO (1961). It is based on ionic hydration. The author postulates that the nucleation of both calcite and aragonite is obstructed by water dipoles adhering to the cations which make up a nucleus. The loosening action of the water molecules upon the nucleus is supposed to be more critical in the case of aragonite where only a rather dense aggregation of ions may become effective as a nucleus. More open aggregations may be sufficient as nuclei for calcite with its lower density, and the water molecules contained in such aggregations are more easily released from the more open structure. The ion-dipole bond is loosened by thermal motion with increasing temperature, and so aragonite nuclei may form with greater ease in hot solutions. GOTO supports his view by the observation that the proportion of aragonite increases at the expense of calcite in precipitation experiments, when the dielectric constant of the aqueous solution is progressively lowered by the addition of diverse alcohols. More data are needed to elaborate the hypothesis of GOTO. If verified, it would be useful in explaining the observation that certain marine organisms, secreting both aragonite and calcite, show an increase in aragonite in warmer waters (LOWENSTAM, 1954).

2. The Influence of Bivalent Cations Other than Ca^{2+}

a) Large Cations, Notably Strontium

The old hypothesis that the formation of aragonite might be favored by the presence of the larger cations Sr, Pb and Ba is based chiefly on the isomorphism of aragonite with strontianite, cerussite, and witherite. Work by CREDNER (1870) is usually quoted in support of this view. The author used the method of CO_2 escape from bicarbonate solutions. However, the data concerning the concentrations used are very scanty, and the criteria used in identifying the precipitates are not altogether convincing. If indeed CREDNER obtained aragonite in the presence of dissolved large cations he did so also in the absence of such additions. As far as the stability of aragonite is concerned, the hypothesis of the influence of Sr, Pb and Ba has been refuted on p. 100, where it was shown that unrealistically high amounts of Sr must be incorporated into aragonite in order to stabilize its structure at normal pressure. However, in view of the structural relationships among the alkaline-earth carbonates, it could be conceived that as cosolutes the larger cations might be conducive to the preferred formation of aragonite nuclei. A paper by ZELLER and WRAY (1956) is usually quoted in support, although it contains no pertinent experimental evidence. On the contrary, the authors show that Sr is incorporated into calcite as evidenced by enlarged cell dimensions. Furthermore, TERADA (1953) has prepared near room-temperature, not only mixed-crystals of calcite type containing 10 mol% $SrCO_3$, but he was able also to incorporate up to 55 mol% $BaCO_3$ into a calcite-type structure. He used the method of adding a rather concentrated (1 N) solution of Na_2CO_3 to 1 N solutions of mixed alkaline-earth nitrates. Similar experiments were carried out by GOTO (1961) who obtained essentially the same results and stated that strontium and barium do not favor the formation of aragonite. It is interesting to note that both TERADA and GOTO observed that moderate additions of barium cause $CaCO_3$ to crystallize as vaterite.

KITANO and coworkers have studied the problem using the method of CO_2 escape from calcium bicarbonate solutions near room-temperature. At first sight, it appears that KITANO and KAWASAKI (1958) obtained indications for a positive influence of strontium additions on the formation of aragonite. However, the positive influence seems to be restricted to an optimum molar ratio of Sr/Ca of 0.002 to 0.003 in solution. This apparently agrees with results of MURRAY (1954) who reported a maximum of aragonite precipitation at molar ratios Sr/Ca from 0.004 to 0.001 in similar experiments. However, in both studies, the formation of calcite could not be entirely suppressed by strontium, not even at its optimum concentration, and increasing amounts of strontium in solution again restored the predominance of calcite. Depending on experimental details, such as stirring or the presence of Cl^- ions, maximum development of aragonite could occur also at the minimum of Sr so that the development of calcite would appear to be favored by Sr. No conclusive evidence regarding a positive influence of Sr on the nucleation of aragonite may be deduced from the studies of MURRAY and of KITANO and KAWASAKI, but the possibility of an optimum Sr/Ca ratio, somewhere below 0.01, favoring aragonite nucleation under certain conditions, cannot be completely dismissed without further investigations.

b) Small Cations, Notably Magnesium

At first sight, a negative influence of magnesium on the formation of aragonite should be expected when the known crystal-chemical relationships are considered. A $MgCO_3$ aragonite-type polymorph does not exist, and most natural aragonites are practically free from magnesium. Magnesium as a cosolute with calcium should be expected to lead to the formation of a rhombohedral structure, and dolomite should appear instead of calcite at higher concentrations of magnesium. However, contrary to such expectations, there is convincing evidence, both experimental and observational, that aragonite crystallizes from aqueous solutions under the influence of dissolved magnesium, provided that its concentration is sufficiently high.

From the observation that aragonite sinter was being deposited at low temperature in the galleries of certain Austrian mines and from very few simple experiments, CORNU (1907) concluded that magnesium sulphate, as a cosolute, gives rise to aragonite precipitation. LEITMEIER (1910, 1916) carried out more experiments in this direction, using the method of CO_2 escape near room-temperature. He found magnesium chloride as effective as the sulphate in forming aragonite. However, these results were subordinated to the view that aragonite precipitates preferably from hot solutions. Perhaps this was because the effect of magnesium on the polymorphism of $CaCO_3$ seemed to contradict isomorphic relationships, and these were about to gain more weight as they were being confirmed by X-ray diffraction. Nevertheless, the effectiveness of magnesium in bringing about aragonite precipitation was corroborated by MURRAY (1954) in tests similar to those of LEITMEIER's.

It has been known, at least since the work of GEE and REVELLE (1932), that aragonite is the product of inorganic carbonate precipitation from sea water. MONAGHAN and LYTLE (1956) precipitated calcium carbonate by adding sodium carbonate to artificial sea waters from which one or more main constituents had been omitted. In the absence of magnesium they obtained calcite, and in the presence of magnesium the product was aragonite (+ monohydrocalcite), no matter which of the other main constituents were present or had been omitted. Later, SIMKISS (1964) obtained similar results using the method of CO_2 escape (see also KITANO and HOOD, 1962).

In precipitation experiments involving additions of dissolved alkali carbonate, instantaneous precipitation usually occurs, due to local supersaturations of unknown magnitude before the solution is completely homogenized. Moreover, the formation of instantaneous precipitates, which are mostly ill-defined (e.g. amorphous) and different from the final product, changes the initial concentration to some unknown degree. Consequently the concentrations characteristic of the final product are not known with certainty. In the method of CO_2 escape from bicarbonate solution, the initial concentrations, at least of the cations, give a more accurate picture of the actual conditions of formation. Nevertheless, the tendency of the precipitate to form at the surface of the liquid, under quiescent conditions, is indicative of considerable concentration gradients. Stirring may be helpful, but it leads again to unrealistically high precipitation rates, comparable to those occurring in the method of alkali carbonate addition.

It is, therefore, important that the tendency of magnesium to provoke the crystallization of aragonite is observed also at the slow rates which are inherent in the method of precipitation from homogeneous solution (cf. GORDON, SALUTSKY and WILLARD). In this method, one of the component ions of the precipitate, usually the anion, is not initially present in solution. Instead, it is generated by the slow decomposition of a suitable dissolved compound. Potassium cyanate, KOCN, decomposes in aqueous solutions, yielding potassium and ammonium carbonate according to:

$$KOCN + 2H_2O \rightarrow K^+ + NH_4^+ + CO_3^{2-} \tag{12}$$

At 100° C the decomposition becomes noticeable in the course of hours, and at room temperature it extends over weeks and months. As early as 1892, BOURGEOIS and TRAUBE used potassium cyanate in attempts to synthesize dolomite at 130° C. Similar experiments at 100° and 20° C (LIPPMANN, 1960) have failed to yield dolomite. Instead, information was obtained concerning the formation of aragonite versus calcite under the influence of magnesium. Series of solutions were prepared containing fixed amounts of Ca^{2+} and KOCN but varying amounts of Mg^{2+}. The sulphates of Ca and Mg were preferred, because they are easier to handle than the deliquescent chlorides. Nevertheless, crucial experiments were carried out also with chlorides, and at different concentrations of Ca and KOCN as well. The results were essentially the same as those summarized in Tables 23 and 24. Calcite is the only product to crystallize in solutions free from magnesium or containing magne-

Table 23. Precipitation of $CaCO_3$ from homogeneous solution at 20° C in the presence of varying amounts of Mg^{2+} (LIPPMANN, 1960).
$CaSO_4 \cdot 2H_2O$:   480 mg/l or 0.00279 M (sea water: 0.01 M Ca^{2+})
KOCN: 1000 mg/l or 0.012 M

$MgSO_4 \cdot 7H_2O$		Molar ratio	Time in days	Phase(s) formed
mg/l	molarity	Mg/Ca		
7000	0.028	10.2	65	A
6000	0.024	8.7	205	A
5000	0.020	7.27	25	A
4000	0.016	5.80	28	A
3000	0.012	4.36	65	A
2000	0.008	2.91	171	A+C
1000	0.004	1.45	53	A
750	0.003	1.09	24	C
500	0.002	0.73	23	C
0	0	0	39	C
(sea water	0.054 M Mg^{2+}	5.1)		

$Na_2SO_4 \cdot 10H_2O$		Molar ratio SO_4/Ca (total)	Time in days	Phase(s) formed
mg/l	molarity			
10000	0.031	12.1	89	C
18000	0.056	21.0	91	C

A = aragonite;   C = calcite.

Table 24. Precipitation of $CaCO_3$ from homogeneous solution at $20°C$ in the presence of varying amounts of Mg^{2+} (LIPPMANN, unpublished)

$CaSO_4 \cdot 2H_2O$:   120 mg/l or 0.0007 M
KOCN: 2000 mg/l or 0.024 M

$MgSO_4 \cdot 7H_2O$		Molar ratio	Time in days	Phase(s) formed
mg/l	molarity	Mg/Ca		
12000	0.048	69	210	A
8000	0.032	46	210	A
6000	0.024	34	210	A
4000	0.016	23	42	A
2000	0.008	11.5	210	A
1000	0.004	5.7	57	C + A
600	0.0024	3.4	36	C
300	0.0012	1.7	42	C
150	0.0006	0.86	57	C

A = aragonite; C = calcite.

sium up to a certain limit. At higher concentrations of magnesium, aragonite appears in addition to calcite, or is the only phase precipitated. LIPPMANN (1960) used the molar ratio Mg/Ca as an *ad hoc* index for the relative efficiency of dissolved magnesium in causing the appearance of aragonite. A ratio Mg/Ca between 3 and 4 appeared to be sufficient for the complete suppression of calcite under the conditions given in Table 23. However, additional data (Table 24) suggest that the critical parameter may be the absolute concentration of magnesium. 0.01 M appears to be the critical order of magnitude above which aragonite forms as the only phase at normal temperature. The possibility that the sulphate ion might be instrumental in the precipitation of aragonite, is disproved by runs in which $MgSO_4 \cdot 7H_2O$ was replaced by $Na_2SO_4 \cdot 10H_2O$ (Table 23).

In view of the slow decomposition of the potassium cyanate at room temperature, one should expect that calcium carbonate would crystallize at a comparable rate. However, in the case of aragonite formation, no change was noticed during several weeks, and then suddenly, sometimes in the course of one day, a substantial amount of precipitate appeared. This is a typical supersaturation phenomenon, which shows that spontaneous nucleation is strongly inhibited for aragonite at ordinary temperature. The occasional formation of small amounts of monohydrocalcite in addition to aragonite in the experiments of Tables 23 and 24 is in line with GOTO's hypothesis (p. 105) that aragonite nucleation is inhibited by the dehydration barrier of the calcium (see p. 88).

The reaction times indicated in Tables 23 and 24 approximately reflect the sudden appearance of an aragonite precipitate. Durations of more than 100 days are no longer characteristic of the time when precipitation took place. When a precipitate failed to appear after about three months, the experiments were no longer inspected daily. Calcite formed before the times given in Tables 23 and 24, but these deposits were identified about parallel with some of the aragonite precipitates.

All of the precipitates adhered to the bottom and the walls of the glass vessels, and aragonite was found to have marked scratches and other irregularities in the glass surface. Aragonite was also observed on such accidental impurities as cellulose fibers introduced from filter paper. Thus the differences in precipitation time experienced for aragonite under comparable conditions appear to be due to varying amounts of inevitable accidental impurities and to slight unintentional differences in the pretreatment of the vessels used. This sensitivity of aragonite precipitation to circumstantial conditions is characteristic of the difficulties opposing the nucleation of the phase at normal temperature.

As indicated on p. 105, when minute amounts of aragonite seeds had been added to a solution free of magnesium which would yield calcite without seeding, aragonite formed as the only phase. This shows that the rate of crystallization for aragonite is greater than that of calcite, although the opposite is true for the nucleation rates at ordinary temperature. Seeding with calcite did not avert the formation of aragonite in runs with magnesium. This observation is an important criterion to be used below in explaining why dissolved magnesium causes the formation of aragonite.

Less potassium cyanate must be used in experiments at 100° C because of the enhanced decomposition rate in order to avoid excessive precipitation rates similar to those obtained by adding alkali carbonate directly. Except for the use of less potassium cyanate, the experiments at 100° C summarized in Table 25 are comparable to those in Table 23 for 20° C. At 100° C, less magnesium is needed, both in absolute concentration and relative to Ca to prevent the formation of calcite. A similar temperature dependence of the efficacy of magnesium in forming aragonite had been observed by LEITMEIER (1910) between 2° and 20° C, and KITANO and HOOD (1962) confirm the same tendency between 10° and 27° C. More work in this direction appears desirable. According to Table 25, magnesium not only subdues calcite, but is even more effective in suppressing vaterite, which may accompany in small amounts the calcite deposited from magnesium-free solutions. The fact that a little aragonite forms with the bulk of calcite shows that under the conditions of the experiments summarized in Table 25 the solubility of aragonite is still exceeded. The formation of aragonite in the absence of Mg is remi-

Table 25. Precipitation of $CaCO_3$ from homogeneous solution at $100°$ C in the presence of varying amounts of Mg^{2+} (Lippmann, 1960)
$CaSO_4 \cdot 2H_2O$: 480 mg/l or 0.00279 M
KOCN: 100 mg/l or 0.0012 M

$MgSO_4 \cdot 7H_2O$		Molar ratio	Time in hours	Phase(s) formed
mg/l	molarity	Mg/Ca		
2000	0.008	2.91	$9\frac{1}{4}$	A
1500	0.006	2.18	$9\frac{1}{2}$	A
1000	0.004	1.45	$5\frac{1}{4}$	A+C
500	0.002	0.73	$4\frac{1}{2}$	A+C
0	0	0	$3\frac{1}{2}$	C(+A)
0	0	0	$8\frac{1}{2}$	C(+A+V)
0	0	0	$3\frac{1}{2}$	C(+A+V)
0	0	0	4	C(+A+V)

$Na_2SO_4 \cdot 10H_2O$		Molar ratio SO_4/Ca (total)	Time in hours	Phase(s) formed
mg/l	molarity			
1000	0.003	2.11	$4\frac{1}{2}$	C(+A+V)
2000	0.006	3.23	$9\frac{1}{2}$	C(+A+V)

A = aragonite;   C = calcite;   V = vaterite.

niscent of the classical precipitation experiments in hot solutions discussed on p. 104f. In order to eliminate aragonite, concentrations must be lower, in the magnesium-free solution than shown in Table 25, and pure calcite may then be obtained at $100°$ C. In harmony with the results at $20°$ C, additions of sodium sulphate likewise fail to yield more aragonite at $100°$ C. Also the inefficiency of sulphate ions in influencing the qualitative results is evident from comparison with a series of experiments at $100°$ C where the chlorides were used instead of the sulphates (Lippmann, 1960, Table 4).

Since aragonite is metastable at ordinary pressure, its formation in simple experiments must be due to kinetic obstacles. However, the rather long durations of the precipitation experiments from homogeneous solutions at ordinary temperature show that the appearance of aragonite under the influence of dissolved magnesium is not due simply to excessive precipitation rates, as is true in the case of magnesium-free solutions at $100°$ C (see p. 105). The experiments with homogeneous solutions demonstrate some specific influence of magnesium above a certain concentration range, or possibly above a certain Mg/Ca ratio. These critical parameters are clearly exceeded in normal sea water ($Mg^{2+} \approx 0.05$ M; Mg/Ca = 5.1); and aragonite is the only phase separating from sea water, not only in slow laboratory experiments (GEE and REVELLE; SIMKISS), but also in nature.

Ooïds are well-known examples among the aragonitic products forming from sea water in warm climates. Their growth forms will be discussed later on p. 116. The inorganic formation of aragonite needles in the sea was disputed by LOWENSTAM (1955) who wanted to derive the needles chiefly from the decay of marine algae. Nevertheless, the inorganic precipitation of aragonite in the sea is well documented, notably in the form of „whitings" (CLOUD, 1962; WELLS and ILLING, 1964). The sudden precipitation of aragonite crystals in a large body of sea water is indicative of considerable preexisting supersaturation. Although this is often caused by the photosynthetic activity of marine plants consuming CO_2 (CLOUD), and although the release of the supersaturation may in certain cases be triggered by organic activity, e.g. by fish stirring up sediment and thus supplying aragonitic nuclei, the overall process is inorganic.

The rule that aragonite is the product of inorganic carbonate precipitation from sea water probably has fewer exceptions than many other rules accepted in the earth sciences. Apparent exceptions, if encountered, merit careful investigation for possible causal explanations.

Aragonite is forming also in non-marine bodies of saline water which contain sufficient magnesium. Examples are the Great Salt Lake, Utah (EARDLEY, 1938) and the Dead Sea (NEEV and EMERY, 1967). In both cases the critical magnesium concentrations are exceeded. Respective characteristic values are: $Mg^{2+} \approx 0.015\,M$; $Mg/Ca \approx 20$ (after CLARKE, 1924) and $Mg^{2+} \approx 1.5\,M$; $Mg/Ca = 4.4$ (after NEEV and EMERY).

Mineralogists know that aragonite occurs typically among the alteration products of rocks rich in magnesium, such as serpentines and basalts (see PALACHE et al.), where it may have originated as a result of hydrothermal activity or weathering. In general, nothing more than speculation is possible concerning the temperature and the exact magnesium content of the solutions involved. Nevertheless, there can be no doubt that the presence of magnesium is a decisive factor in such occurrences of aragonite. That bivalent iron may act in the same way, by virtue of its similar ionic size, is suggested by the occurrence of aragonite as a supergene product in the siderite operations of the Erzberg near Eisenerz, Austria. Here the mineral is found in the peculiar form of "Eisenblüte" ("flowers of iron") consisting of irregularly curved aragonite stalactites.

As indicated above, the property of magnesium to cause the crystallization of calcium carbonate as aragonite is a paradox from the purely crystal-chemical viewpoint. A rational explanation is possible though, when not only the structures of the solid carbonates are considered, but also the state of the pertinent cations in solution, i.e. when account is taken of cation hydration. In the discussion of the hydrous magnesium carbonates (pp. 71—87), the failure to synthesize magnesite at ordinary temperature was attributed to the firm bonding of water molecules to the magnesium ion. Similar reasoning will explain also the formation of aragonite in the presence of dissolved magnesium.

Even if no information were available concerning the solubility of dolomite, the mere existence of the solubility product, $[Ca^{2+}] \cdot [Mg^{2+}] \cdot [CO_3^{2-}]^2 = k_{sD}$ (see p. 156), indicates that dolomite must become stable relative to calcite as soon as a certain (molar) Mg/Ca ratio is exceeded in solution. Based on the structural similarity of calcite and dolomite, one may guess that this critical ratio is not too far from unity, at least in order of magnitude (cf. Fig. 50). Thus, with increasing Mg/Ca ratio, dolomite should replace calcite in such series of experiments as those described in Tables 23 and 24. Instead, calcite persists and is replaced by aragonite at higher Mg/Ca ratios. It is true that in the experiments just mentioned calcite becomes slightly magnesian with increasing Mg^{2+} in solution, but a content of 5 mol% $MgCO_3$ in the calcites formed was not exceeded at 20° C, judging from the displacement of the strongest calcite reflection. This shows that the adsorption equilibrium for Mg^{2+} versus Ca^{2+} with respect to the rhombohedral phase is more in favor of Ca^{2+}. Even for the partial dehydration concomitant with the adsorption of Mg^{2+} on the crystal surface, the energy barrier appears to be so much greater than for Ca^{2+} that Ca^{2+} is incorporated in excess over the Mg/Ca ratio in solution. If more magnesium is forced upon the surface of the rhombohedral phase by increasingly higher concentrations of dissolved Mg^{2+}, a point may well be reached where half the cation sites in the surface are occupied by magnesium as would be required for the growth of dolomite, but this point would probably pertain to the stability field of magnesite. The actual failure of dolomite to grow in the experiments of Tables 23 and 24 is reminiscent of the practical impossibility of obtaining magnesite under similar conditions even though its thermodynamic solubility is exceeded.

Under B.III.2., a detailed explanation of the difficulties opposing the crystallization of magnesite has been offered. This interpretation applies equally well to other cases where magnesium has to be incorporated into anhydrous crystal structures. Modifications of Fig. 32 may be conceived where only part of the cation sites are occupied by magnesium, the others by calcium, e.g. in a 1:1 ratio for dolomite, or with less magnesium for magnesian calcites. In these cases also, the magnesium ions at the surface are as firmly bonded by water dipoles extending outward as in a magnesite nucleus. Below a certain surface concentration of the magnesium ion, it may become dehydrated as its water molecules are supplanted by CO_3^{2-} ions which are first captured by adjacent surface calcium ions, whose dehydration is easier because of lower interaction energies (Tables 20 and 21). Obviously, this process becomes less feasible with increasing surface concentration of magnesium. Under the conditions depicted in Tables 23 and 24, the growth rate of nuclei with more than 5 mol% $MgCO_3$ is practically imperceptible as is the case with magnesite, because of surface poisoning by firmly hydrated magnesium ions. Continued increase of $[CO_3^{2-}]$ in solution, e.g. by reaction (12) or by CO_2 escape, results in the solubility, or rather the critical supersaturation, for aragonite being exceeded, and nucleation of aragonite takes place followed by precipitation.

One might object that the surface of aragonite nuclei must be blocked as well by adsorbed hydrated magnesium. However, in the discussion of the aragonite structure (p. 63 f.) the point was made that cations smaller than calcium are obviously unable to occupy the symmetric position with respect to the CO_3 groups as required in the aragonite pattern (Fig. 27). Thus, if magnesium is at all adsorbed by the surface of aragonite, it will probably by bonded between two oxygens belonging to two different CO_3 groups (cf. Fig. 45, IV). From this position, it will easily be expelled by the electrostatic repulsion exerted by adjacent calcium ions. These must be expected to be more firmly adsorbed, at least by three oxygens of two CO_3 groups, in harmony with the aragonite pattern. In contrast to calcite-type nuclei poisoned by hydrated magnesium, aragonite may grow freely as long as the supply of Ca^{2+} and CO_3^{2-} exceeds its solubility. No serious kinetic obstacle is created by the presence of magnesium because these ions need not be dehydrated, for they are not incorporated into the aragonite structure. For the same reason, the presence of magnesium is no obstacle to the nucleation of aragonite, which process appears to be retarded rather by the dehydration barrier of the calcium itself (p. 105 f.; GOTO, 1961).

In the explanation given above, the presence of calcite-type nuclei is assumed and the reasoning is independent of whether these have formed spontaneously or have been added to the system from outside. In a sufficiently clean artificial system, it is possible that even the generation of effective nuclei is suppressed by the incorporation of hydrated magnesium ions into aggregations which might otherwise act as nuclei for a rhombohedral carbonate. Nevertheless, the same explanation given above remains valid, since no assumption concerning the size of the nucleus is necessary. Under natural conditions, say in sea water, calcitic nuclei are usually present in the form of organic debris. In particular, the explanation is in line with the observation stated above, that seeds of calcite do not grow in solutions containing sufficient magnesium and

8*

that aragonite appears just as when calcitic seeds are absent. The possibility that aragonite may nucleate faster on calcite seeds than in their absence does not change the overall result. This is the same also in sea water where aragonite is the product of inorganic carbonate precipitation despite the ubiquity of calcitic seeds.

3. Interpretation of Different Growth Forms of Aragonite Oöids

In contrast with ancient oölites, in which the oöids now consist of calcite, recent oöids may be adduced as typical examples for the segregation of aragonite from magnesium-bearing solutions, notably sea water. Under the microscope, between crossed polars, sectioned oöids often exhibit a "pseudo-uniaxial" black cross (BREWSTER's cross) indicating the preferred orientation of the aragonite crystallites with respect to the (near-)spherical shape of the oöid. By taking account of the polarization phenomena which arise on inserting a quartz wedge (SORBY, 1879) or some other compensator, the orientation of the main direction n_α, i.e. the c axis, of the crystallites may be determined. Two different types of preferred orientation have been recognized. In the majority of marine oöids, notably in those from the Bahamas, the c axes of the aragonite crystallites are aligned tangentially with respect to the surface (SORBY, 1879; ILLING, 1954; NEWELL, PURDY and IMBRIE, 1960; RUSNAK, 1960; PURDY, 1963; BATHURST, 1967a, 1968). Other recent oöids, especially those forming in the Great Salt Lake, Utah, show a radial orientation of acicular aragonite crystals elongated in the c direction (EARDLEY, 1938; RUSNAK, 1960). This fabric may be referred to simply as "radial-fibrous" (cf. BATHURST, 1968).

Concentric layering occurs in both radial-fibrous (EARDLEY) and tangentially built oöids. The layering appears to be independent of the fabric type. The layering is marked by changes in grain size, and sometimes by varying amounts of occluded substances, such as unoriented carbonate, organic matter or clay minerals. For the recent oölite of the Laguna Madre, Texas, RUSNAK (1960) has shown that different types of aragonite orientation may occur in the same oöid. He found concentric layers of radially oriented aragonite intercalated between layers consisting of tangentially oriented or unoriented aragonite. The unoriented material may have originated in the same way as the cryptocristalline aragonite cementing grapestones (PURDY, 1963; BATHURST, 1967b) and forming the micrite envelopes around shell fragments (BATHURST, 1966) in recent Bahaman sediments. In both cases, the presence of abundant algae suggests that the CO_2 loss concomitant with their photosynthetic activity may have lead to the precipitation of the micritic aragonite.

In contrast, there is no evidence suggesting a direct organic influence in the deposition of oriented aragonitic oöid fabrics with either radial-fibrous or tangential alignment of the c axes. The view that these fabrics are essentially of inorganic origin is supported by two facts: (1) Tangential orientation of aragonite c axes occurs in the pisolites (Erbsen- oder Sprudelsteine) (SORBY, 1879) which were once deposited from the hot springs of Karlsbad (Karlovy Vary), thus obviously without any organic interference. (2) The aragonite precipitated in laboratory experiments often consists of spherulites with a radial-fibrous arrangement of the c axes. LINCK (1903) appears to have been the first to relate such artificial spherulites to oöids as did later investigators (e.g. MONAGHAN and LYTLE, 1956; OPPENHEIMER, 1961; USDOWSKI, 1963).

Radial-fibrous oöid formation is thus correctly classified under the heading of spherulitic crystal growth. Although this is a well-known phenomenon, occurring in the crystallization of many other minerals and chemicals, only certain geometric aspects of the process are understood so far. The rate of crystallization must be greater in one crystallographic direction than in others for fibrous growth to take place. When crystallization starts from a center, or also from the surface of a preexisting grain of finite dimensions, only those fibrous nuclei which are directed outward in a radial fashion will survive as crystal growth continues. The component of the growth-velocity vectors directed outward has its maximum in these radial fibers. Fibers making an angle with the radial direction have little chance to develop since the space into which they might grow will be occupied by the faster radial fibers. This principle of selective growth of favorably oriented nuclei ("Keimauslese"; GROSS and MÖLLER, 1923; SPANGENBERG, 1935; CORRENS, pp. 164–165) will be used as a lemma in an attempt to elucidate the conditions leading to radial-fibrous growth in aragonite oöids.

The most discussed explanation for the origin of tangential aragonite orientation has been SORBY's (1879) snowball hypothesis: Loose needles of aragonite, elongate in the c direction, are presumed to adhere mechanically to a preexisting grain acting as a nucleus, where they become fixed by further crystallization. Recently, BATHURST (1968) has shown that SORBY's hypothesis cannot be reconciled with certain properties exhibited by the tangentially built laminae in oöids, e.g. with their resistance to abrasion and their strictly monomineralic (aragonitic) character.

As indicated on p. 64, the two different types of preferred orientation suggest a correlation with important features of the aragonite structure. Here strings of calcium ions, not separated by CO_3, are found both in the basal (a, b) plane and also in the perpendicular direction, i.e. along the c axis. In aggregates of calcite showing preferred orientation, only the radial-fibrous type of c-axis orientation is known, e.g. in cave pearls

(FÜCHTBAUER and MÜLLER, p. 328) and in the artificial "oöids" forming in water-softening plants (KNATZ, 1965, 1966, and personal communication). This orientation suggests a correlation with the monoionic lattice rows in the basal plane of calcite, whereas the absence of a tangential c axis fabric for calcite may be related to the absence of monoionic lattice rows along the c axis of the calcite structure.

A different type of evidence, which may help in explaining the two types of aragonite orientation in oöids, comes from precipitation experiments yielding radial-fibrous aragonite spherulites as the final product. LINCK (1903) noticed that artificial spherulites of aragonite frequently occur in pairs grown together. He found that these twin spherulites develop from aragonite crystals elongate in the direction of n_α, i.e. of the c axis. During the intermediate stages of growth, broom-like excrescences formed on either end of the crystals. The resulting shapes reminded LINCK of the sheaf-like crystal aggregates characteristic of stilbite ("desminbündelartige, faserige Aggregate") and he found them related to the twin spherulites by all kinds of transitional developments. An excellent microphotograph showing such sheaf-like aggregates of artificial aragonite has been published by OPPENHEIMER (1961). In the experiments summarized in Tables 23 and 24, the present writer found exactly the same evolution as described by LINCK in the aragonite obtained. Three typical stages are shown in Fig. 40.

In the experiments of Tables 23 and 24 the aragonite begins to form at the initial calcium concentrations stated in the headings of the tables.

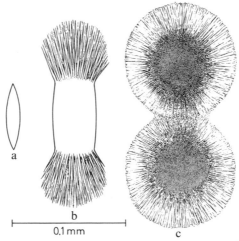

0.1 mm

Fig. 40. Development of aragonite in the experiments summarized in Tables 23 and 24; from needle-like crystals (*a*), via sheaf-like aggregates (*b*), to radial-fibrous twin aggregates (*c*) [see also LINCK (1903) and OPPENHEIMER (1961)]

These concentrations are high in relation to the solubility product of aragonite. In view of what is known concerning the precipitation of other sparingly soluble salts, notably silver halides, it must be concluded that during the early stages of growth, the surfaces of the aragonite crystals are covered with calcium ions. As the crystallization continues, the solution, being a closed system, is depleted in calcium and more CO_3^{2-} is needed to exceed the critical ionic product $[Ca^{2+}] \cdot [CO_3^{2-}]$.

In this manner, a limit will be reached beyond which the aragonite surface becomes covered preferentially with CO_3^{2-} ions.

In order to see this change in surface conditions more clearly, it is advisable to consider information obtained from the silver halides. Consider that a silver halide is precipitated by mixing solutions of silver nitrate and alkali halide. When an excess of silver nitrate is supplied, it is found that the particles precipitated have a positive charge. The reverse occurs with an excess of alkali halide. Silver halide with a negative charge is obtained. Electrophoresis is the usual method for determining the sign of the charge, i.e. the sense of migration of the particles is observed in an electric field. The charge itself is due to preferred adsorption of the excess ion, either Ag^+ or e.g. Cl^-, on the surface of the silver halide crystallites. It must be assumed that the adsorbed excess ions mingle with the other surface ions and are indistinguishable from these (OVERBEEK, 1952; ADAMSON, 1960). Discounting surface imperfection, the adsorbed ions causing the surface charge will occupy the regular cation or anion sites of the crystal structure, e.g. of an NaCl pattern in AgCl and AgBr. In textbooks of colloid chemistry, AgI is commonly chosen to illustrate the relationship between surface charge and solution composition (e.g. JIRGENSONS and STRAUMANIS, pp. 96, 148). However, with silver iodide the situation is complicated by the occurrence of several polymorphs (BURLEY, 1963). It is, therefore, safer to consider silver chloride and bromide which precipitate only as NaCl structures.

In general, zero surface charge does not coincide with equivalent concentrations of cations and anions in solution. Under the latter condition, the silver halides show a negative charge, and a slight excess of dissolved Ag^+ is needed to obtain electrically neutral particles. This shows that the halide ions are more tightly bonded to the crystal surface than the smaller silver ions. These have a more pronounced tendency to be hydrated and thus to occur in solution. The concentrations characterizing zero charge may be found in OVERBEEK (1952, pp. 231—232). For example, AgCl has $[Ag^+] = 10^{-4}$ and $[Cl^-] = 10^{-5.7}$.

Similar data derived from electrophoresis do not seem to be available for precipitated calcium carbonates, probably because their solubility is too great and the crystals obtained are too coarse-grained for that method. Nevertheless, the calcium carbonates may be expected to show a qualitative behavior similar to the silver halides. In particular, given the hydration properties of the ions involved (Table 20), zero charge should occur at a slight excess of dissolved calcium. This is borne out by the electroosmotic experiments which DOUGLAS and WALKER (1950) carried out using ground calcite. In pure water they observed a negative surface charge, which indicates a preferred release of calcium into solution, resulting in a surface coating of CO_3^{2-} ions. Zero charge was obtained in a 0.0004 M solution of calcium chloride, and greater concentrations gave rise to a positive charge. DOUGLAS and WALKER also studied the influence of other ions on the surface charge of calcite. They found that bivalent cations other than calcium had a similar effect, but were less efficient in shifting the charge in the positive direction. Multivalent anions, notably carbonate, and to a lesser extent sulphate, caused the surface charge to be more negative than in pure water.

Thus qualitatively, calcite shows the same dependence of the surface charge as the silver halides. In spite of the absence of experimental data for aragonite, it is sound to assume an analogous behavior. Although the calcium concentration leading to zero charge can be different, the order of magnitude may be expected to be similar to that for calcite. The calcium concentrations causing zero charge will vary anyway in natural systems, due to the presence of other potential-determining ions, such as Mg^{2+}, CO_3^{2-} and SO_4^{2-}. Nevertheless, in a given system, the surface charge of, say aragonite, will depend essentially on $[Ca^{2+}]$ and $[CO_3^{2-}]$. Since these concentrations are interdependent via the critical ionic product $[Ca^{2+}] \cdot [CO_3^{2-}]$ it will be sufficient in qualitative explanations to consider only one concentration, e.g. $[Ca^{2+}]$. It must also be borne in mind that the critical ionic product which must be exceeded for crystal growth to proceed is usually greater (due to supersaturation) than the solubility product and that it may be different for different growth forms.

From analogy with the silver halides, it may be concluded that the adsorbed excess ions are indistinguishable from the other ions making up the crystal surface of a calcium carbonate. However, the conclusion that the surface ions also occupy the same sites as their counterparts inside the crystal is strictly valid only for the calcium ions, the situation being different for the CO_3 groups. These not only occupy certain sites in a crystal structure, but in addition they also occur in certain orientations, i.e. in a coplanar alignment in calcite and aragonite. In contrast, the CO_3^{2-} ions adsorbed on the surface are free to assume all kinds of orientations, or misorientations, with respect to the CO_3 groups inside the crystal, since they are not buttressed by calcium ions from all sides. This is true in particular when the calcium concentration in solution is so low (or the CO_3^{2-} concentration so high) that a negative surface charge exists, i.e. when an aragonite nucleus is coated preferentially by

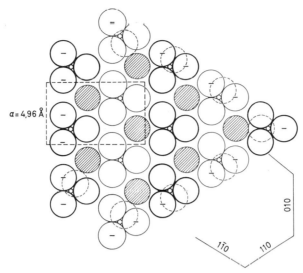

Fig. 41. A nucleus of aragonite with excess CO_3^{2-} in the surface. Projection along the c axis on the basal plane (a, b). The dashed rectangle is the (a, b) outline of a unit cell (cf. Fig. 27). The outline at the lower right shows the most common morphological faces. – The surface CO_3^{2-} ions are shown as full circles in the ideal orientation which they would have inside the crystal. It must be expected, however, that they are actually misoriented by tilts of various degrees, in particular through the electrostatic repulsion of O^{x-} ions belonging to different CO_3^{2-} groups. Possibilities for maximum tilt are suggested by dashed circles which represent two oxygens on top of each other. The CO_3 tilts here indicated are perhaps too schematic because only one Ca level (shaded circles) and parts of two adjacent CO_3 layers are shown. The mutual interference among CO_3^{2-} belonging to different layers is thus not apparent in this figure, but see Fig. 42

CO_3^{2-} ions. Fig. 41 shows various possibilities of CO_3^{2-} tilt in an aragonite nucleus covered by CO_3^{2-}.

When a Ca^{2+} is captured by such a nucleus it will be immediately followed by a CO_3^{2-} ion restoring the negative surface charge (Fig. 42). The CO_3^{2-} ions which were originally exposed at the surface and which have captured the Ca^{2+} have thus become incorporated into the crystal. It may be conceived that in this act, at least some of the CO_3 groups fail to assume the exact orientation required for an ideal aragonite crystal.

The misoriented CO_3 groups will cause the crystal to grow in directions which deviate from the orientation of the original nucleus. Successive misorientations may add up to compound deviations of 90° and more. Since the principle of "Keimauslese" (selective growth of favorably oriented nuclei, see above) is active at the same time, radial-fibrous aggregates of aragonite will be the result.

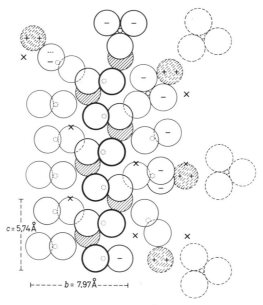

Fig. 42. A nucleus of aragonite with excess CO_3^{2-} in the surface (right side and top). – Projection along the a axis on the (b, c) plane. Due to the misorientation of the surface CO_3^{2-}, incoming Ca^{2+} ions (dashed shaded circles) are captured in sites which differ from the regular Ca sites marked by crosses ×. The immediately following CO_3^{2-} (dashed), which restores the negative surface charge, will thus be incorporated with still greater misorientation (here exaggerated). The presumed effect is crystal growth deviating in orientation from the original nucleus and thus resulting in radial-fibrous aggregates

The explanation just given implies that spherulites should form preferably in such compounds which contain complex atomic groups susceptible of misalignment. Although a survey in this direction of minerals and other inorganic compounds is not yet available, the overall impression seems to be in favor of such a view. Furthermore, the interpretation might be verified by testing whether radial-fibrous growth occurs indeed when the complex ion is present in solution in excess over the simple component ion(s). At any rate, the hypothesis is in harmony with the experience that radial-fibrous growth is frequently observed in the crystallization of organic compounds.

It was stated already that the development of aragonite as shown in Fig. 40 presumably started out with Ca^{2+}-coated nuclei growing into aragonite needles (Fig. 40a). According to the explanation for radial-fibrous growth, the appearance of the broom-shaped excrescences (Fig. 40b) must mark the change to a surface covered by CO_3^{2-}, the change being caused by the consumption of dissolved Ca^{2+} due to the crystallization of aragonite. Increasing misalignment of the excrescences, checked though by "Keimauslese", finally leads to a twin spherulite as shown in Fig. 40 c.

In nature, radial-fibrous aragonite growth usually starts from a preexisting detrital particle acting as a nucleus. In a solution poor in Ca^{2+}, CO_3^{2-} will be adsorbed on the particle in various orientations. Crystal growth in all directions should thus be possible. However, directional selectivity will allow aragonite fibers to grow only in the radial direction.

The molarity of calcium (0.01 M) in sea-water appears to be too high for radial-fibrous growth to take place in the open system. Nevertheless, radial-fibrous layers occur in the oöids of the Laguna Madre (RUSNAK, 1960) and as overgrowths on oöids buried in the sediment near the Bahamas (BALL, 1967). From such occurrences it may be inferred that a slight decrease of dissolved calcium, due to aragonite precipitation, is sufficient to cause the reversal from Ca^{2+}-covered to CO_3^{2-}-covered nuclei in a more or less closed system. According to analyses quoted by CLARKE (p. 157), calcium may be considerably lower in the Great Salt Lake than in sea water, and this should be the reason for the formation of radial-fibrous oöids in that locality.

Normal sea water (0.01 M Ca) or solutions still higher in Ca^{2+}, e.g. the brine of the Dead Sea (~ 0.34 M) are characterized by the spontaneous precipitation of free aragonite needles (CLOUD, 1962; WELLS and ILLING, 1964; or NEEV and EMERY, 1967; respectively).

An aragonite nucleus covered by Ca^{2+} ions is shown in Figs. 43 and 44. In this configuration, an incoming CO_3^{2-} ion will find its site clearly defined and there is little chance for misorientation since the CO_3^{2-} is held in position by Ca^{2+} ions already part of the crystal and by the immediately subsequent Ca^{2+} ion which restores the positive surface charge. Thus crystal growth will proceed strictly in the orientation prescribed by the nucleus. Normal crystals, e.g. aragonite needles, will be the result, as long as the concentration is not essentially lowered by the crystallization itself.

When precipitating in the form of "whitings", aragonite needles originate essentially by homogeneous nucleation, at least according to the observations of WELLS and ILLING. This nucleation mechanism is known to require extreme supersaturations. It is also known that sea water supersaturated with aragonite may exist without precipitation, i.e. homogeneous nucleation, taking place. The supersaturation is released only when the water is brought into contact with appropriate nuclei (see e.g. WEYL, 1961). It seems likely that tangentially oriented aragonitic layers grow on oöids under similar conditions (BATHURST, 1967b). The close

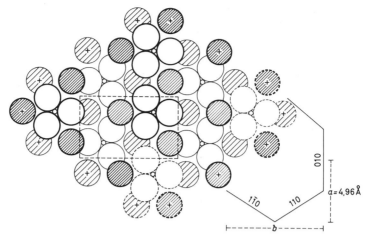

Fig. 43. A nucleus of aragonite with excess Ca^{2+} adsorbed on the surface. Projection along the c axis on the basal plane (a, b). Incoming CO_3^{2-} ions (dashed) assume the correct orientation under the influence of Ca^{2+} adsorbed already (full, shaded circles) and Ca^{2+} (dashed) immediately following to restore the positive surface charge

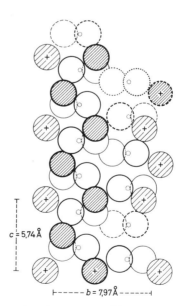

Fig. 44. A nucleus of aragonite with excess Ca^{2+} adsorbed on the surface. Projection along the a axis on the (b, c) plane. Captured CO_3^{2-} (dashed, dotted) are correctly orientated by sufficient Ca^{2+} (see also Fig. 43). Undisturbed crystals, e.g. straight needles, are the result of growth under these conditions, which are presumed to apply also to the formation of tangentially built oöids

association of oölitic sediment and aragonite-needle mud in the Bahamas (BATHURST, 1968) suggests indeed that both types of sediment form
from similar waters, and a difference can be seen chiefly in the degree of
supersaturation. A lower supersaturation than is required for the spontaneous nucleation of aragonite crystals will be sufficient to promote the
growth of oölitic layers on preexisting nuclei.

In open sea water a detrital grain acting as a nucleus for oölitic
growth will have its surface covered by Ca^{2+} ions. Among these, pairs of
Ca^{2+} ions will exist which have nearly the same distance as in the crystal
structure of aragonite (Fig. 45, I). Such pairs of Ca^{2+} should be the preferred sites for the initiation of heterogeneous nucleation. They should
be capable of capturing a CO_3^{2-} in about the same way as do the cations
in the surface of a crystal proper. At first sight, it would appear that the
resulting tangential alignment of the CO_3^{2-} (Fig. 45, II) might give rise to
a radial orientation of the c axes of aragonite. However, the Ca^{2+} which
immediately follows the CO_3^{2-} to restore the positive surface charge will
rotate the CO_3 group by about $90°$ (Fig. 45, II—III) by means of the
electrostatic repulsion between this last Ca^{2+} ion and the first two Ca^{2+}

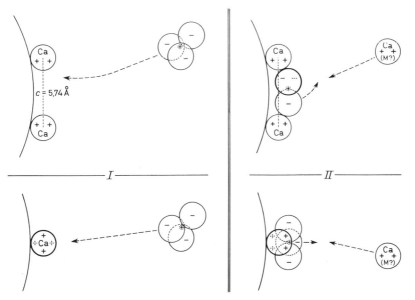

Fig. 45. Proposed formation of a (c-axis) tangential nucleus of aragonite on a preexisting (near) spherical substratum, e.g. a detrital grain, endowed with a Ca^{2+}
surface layer. Lower parts: plan; upper parts: elevation. Two Ca^{2+} ions, either
part of the substratum or adsorbed to it, happen to be separated (approximately)
by the c spacing of aragonite (stage I). They capture a CO_3^{2-} ion (stage I–II).
A third cation which follows restores the positive surface charge (stage II)

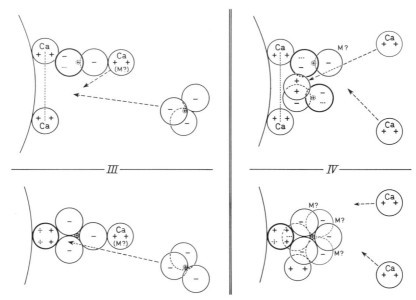

Fig. 45 (continued). By the addition of the third cation, the CO_3 group has rotated to radial orientation, and the tangential c axis alignment of the nucleus is now secured (stage III). The third cation (stage II–III) is marked Ca(M?) since it might be magnesium instead of calcium. Because a magnesium ion cannot occupy the site of the third fixed cation in stage IV, the short arrow in stage III (elevation) does not apply in the case of magnesium, and the third calcium of the nucleus in stage IV must be captured from solution. – Possible sites of magnesium adsorbed between two oxygens (belonging to different CO_3^{2-} groups), or to one oxygen, are marked by M?. However, it is possible that Mg^{2+} is not at all adsorbed on an aragonite surface

ions (and perhaps other Ca^{2+} adsorbed on the grain). The resulting configuration (stage III) permits the addition of another CO_3^{2-} ion. This process is accompanied by a change in position of the preceding Ca^{2+} (stages III-IV). The configuration of stage IV offers at least two sites suitable for the adsorption of Ca^{2+} which maintain the positive surface charge. After the third CO_3^{2-} has been captured (stages V—VI) there are several sites available for Ca^{2+} addition. They are marked by crosses $\times$ in stage VI.

At this point, the illustration of the heterogeneous aragonite nucleation on a Ca^{2+}-coated detrital substratum is discontinued. On the one hand, it would be prohibitive to follow by illustrations the several possibilities for the further development of the nucleus which ensue from stage VI, and it would be somewhat arbitrary to follow only one possibility. On the other hand, as far as the orientation problem is concerned,

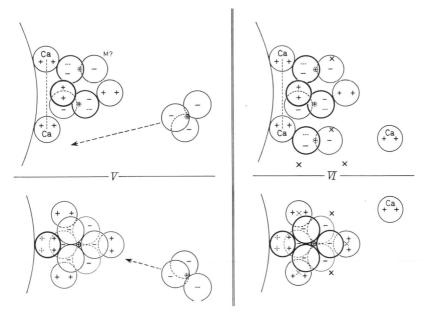

Fig. 45 (continued). More component ions are added to the nucleus in stage V. The several sites available for the addition of Ca^{2+} in stage VI are marked by crosses ×

it would be pointless to follow the further growth of the nucleus, since the rotation of the CO_3 to radial alignment has already taken place between the stages II and III; and this rotation is presumably the decisive step which gives rise to the tangential orientation of the c axis with respect to the surface of the preexisting grain.

Fig. 45 gives the impression that the nucleation mechanism shown results not only in the tangential orientation of the c axis but also in the additional radial orientation of the b axis of aragonite. For the sake of symmetry, the b axis was indeed chosen normal (and the a axis parallel) to the preexisting grain. However, as long as the nucleus is fixed on the latter only by the first two Ca^{2+}, rotation of the nucleus about the axis connecting these two Ca^{2+} is possible within limits. For example, the third Ca^{2+} (stage IV) may adhere to a favorable site of the substratum. Moreover, if the diagonal of the (a, b) outline, i.e. the [110] direction of aragonite, were chosen parallel to the substratum, the resulting configuration would be closely similar to that actually shown in Fig. 45, since [110] and [100] (= the a axis) are related by the (110) twinning so common in aragonite (see Fig. 28). Thus the proposed nucleation mechanism yields nothing more than a preferred tangential orientation of the c axis, whereas the b (and a) axes are randomly oriented. This appears to be in harmony with observational evidence (e.g. PURDY, 1963).

In the sketches making up Fig. 45, the first two Ca^{2+} of the nucleus are supposed to be adsorbed on an exactly spherical, smooth grain. It is easy to see that c axis orientations deviating from the exactly tangential alignment may occur on less regular surfaces as are characteristic of natural detrital grains. However, also in the case of nearly tangential crystal growth, a number of factors make for directional selectivity, though not in an equally stringent fashion as in the case of radial-fibrous growth. These factors are: (1) A fiber (or plate), making even a small angle with the surface, can grow only in one direction, i.e. outward. A nucleus growing exactly in the tangential direction, at the same rate, will gain about double the mass during a given time, since it propagates in two directions. (2) The most efficient way of filling the space in the angle between a misoriented crystal and the surface of the substratum is by crystals growing in the tangential direction. (3) Crystals whose main growth direction deviates from tangential will protrude more than those showing the exactly tangential alignment. Thus the former are exposed more to abrasion.

By the explanation of aragonite nucleation on a Ca^{2+}-covered surface (as illustrated in Fig. 45, I—VI), the tangential orientation of the c axis is now clearly correlated with the monoionic lattice rows along the c axis of the aragonite structure. Furthermore, the non-existence of calcitic fabrics with a tangential c-axis orientation now follows automatically from the absence of corresponding monoionic lattice rows in the calcite structure. When the nucleation of calcite on a Ca^{2+}-coated substratum is envisaged, it may be expected that the first CO_3^{2-} ion is captured by three Ca^{2+} ions forming an equilateral triangle similar to the hexagonal base of calcite (see Fig. 4). The resulting nucleus (Fig. 46) has its c axis perpendicular to the substratum, and the simultaneous growth of many such nuclei will yield the radial c-axis orientation of calcite as it is known from stalactites, stalagmites, and cave-pearls. Since most of the nuclei start out with the right orientation, the corrections to be effected by directional selectivity ("Keimauslese") are not as essential as they are in the case of the radial growth of aragonite. This has been postulated above to start from completely random orientation to be ascribed to misoriented CO_3^{2-} ions on the surface of an aragonite nucleus (see Fig. 42).

One may now enquire into what will happen to a calcitic nucleus with a CO_3^{2-} surface population imparted by high dissolved $[CO_3^{2-}]$. Due to the imbricate arrangement of the component ions in calcite, a moderate excess of CO_3^{2-} in the surface will not give rise to misorientation (Fig. 47, upper part). But it is doubtful whether a calcite nucleus with

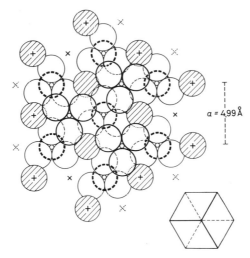

Fig. 46. A nucleus of calcite with excess Ca^{2+} adsorbed on the surface. Projection along the c_{hex} axis on the basal plane (cf. Fig. 4). Of the sites suitable for the addition of CO_3^{2-}, the C centers are marked by crosses $\times$. Possible sites for Ca^{2+} restoring the positive surface charge are shown as dashed, heavy circles. As in the case of aragonite under the same conditions (Fig. 43 and 44), CO_3^{2-} will be captured in the correct orientation by the influence of the Ca^{2+} ions. – The orientation of the morphologic rhombohedron is shown on the lower right

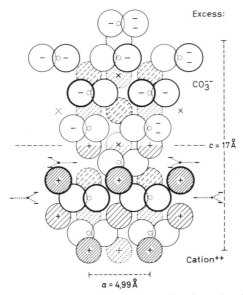

Fig. 47. A nucleus of calcite elongate in the c_{hex} direction, showing surface excess of either Ca^{2+} (lower part) or CO_3^{2-} (upper part)

9 Lippmann

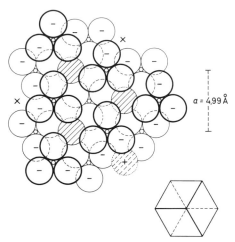

Fig. 48. A hypothetical nucleus of calcite with an excess of adsorbed CO_3^{2-}. – Projection along the c_{hex} axis on the basal plane. It is doubtful whether the surface CO_3^{2-} can maintain the ideal orientation here shown. Under the influence of the electrostatic repulsion among the surface CO_3^{2-}, misorientations must be expected of such an extent that the surface pattern is no longer similar to a calcite structure

a high surface excess of CO_3^{2-} may exist as shown in Fig. 48. Under the influence of electrostatic repulsion the CO_3^{2-} ions should show orientations different from Fig. 48. However, a cooperative CO_3^{2-} tilt of the type which may lead to greater O–O distances among adjacent CO_3^{2-} ions in the surface of aragonite (Fig. 42) does not appear possible in calcite without profound changes of the surface pattern. Misoriented CO_3^{2-} would affect the calcite surface to such an extent that the resulting pattern would have little resemblance with the calcite structure, and the growth of ordered calcite must be strongly inhibited by a surface so poorly ordered. It may be tentatively assumed that vaterite forms more easily under these conditions, i.e. from solutions high in CO_3^{2-}. For the rather open structure of vaterite will allow greater O–O separations among adjacent CO_3^{2-} groups. In addition, the disorder phenomena so common in vaterite attest to considerable mobility of the CO_3 groups in this structure, and this should *a fortiori* apply to a growing surface. The view that a high concentration of CO_3^{2-} leads to vaterite is not at variance with experience. Nevertheless, additional experiments specially designed to test this hypothesis appear highly desirable.

The rôle of magnesium in the formation of aragonite has been amply explained on pp. 113—115. According to the interpretation given there, magnesium does not enter directly into the precipitation of aragonite, but it appears to act indirectly in that it suppresses the crystallization of calcite (and vaterite). Therefore, it was not deemed necessary to mention the influence of the magnesium in the structural interpretation of aragonite growth forms, although dissolved magnesium is needed to check the formation of calcite. Nevertheless, in view of the possibility that some magnesium may perhaps be adsorbed also on the surface of aragonite, sites suitable for magnesium are indicated by M? in some of the stages of aragonite nucleation sketched in Fig. 45. For example, the cation effecting the rotation of the CO_3 group (Fig. 45, II—III) might well be magnesium. In this case, the third Ca^{2+} present in the nucleus in stage IV must have become adsorbed immediately after the second CO_3^{2-}. The further development of the nucleus will then take a course slightly different from what is shown in Fig. 45, IV—VI, unless the Mg^{2+} in question is lost again at stage IV. Other sites suitable for Mg^{2+} adsorption are shown in stage IV. They are between two oxygens not in the same CO_3 group and thus different from Ca^{2+} sites. Any Mg^{2+} occupying (part of) these sites will not obstruct the addition of the third CO_3^{2-} and will be expelled by following Ca^{2+}.

The preceding discussion may appear highly speculative. Even so, it is soundly based on crystal structure and experience from colloid (or surface) chemistry. It is quite possible that some of the views exposed above may need modification as soon as more experimental data become available. At the same time, it is felt that the structural approach may also lead to new experiments helping to elucidate the conditions of formation for the various growth forms occurring in the carbonates.

4. The Formation of Calcite and Aragonite in Organisms

It is well known that organisms secreting calcium carbonate as their hard parts may contain calcite, or aragonite, or both. The most reliable method now available for the determination of skeletal carbonates is X-ray diffraction. MAYER and WEINECK (1932) were the first to make extensive use of this method and they essentially confirmed older data derived from density measurements, staining tests, and microscopic observations. Since then, more X-ray determinations of skeletal carbonates have accumulated, notably through the work of CHAVE (1954) and LOWENSTAM (1954). Some of these results are summarized in Table 13. More data may be found in a paper by DODD (1967).

There is, however, no generally accepted hypothesis as to what may cause organisms to deposit one polymorph, or the other, or both. In view of the distinct determining influence of dissolved magnesium on the polymorphism of calcium carbonate as demonstrated and explained on pp. 107—115, the question naturally suggests itself as to whether magnesium is the determining factor also in organisms. This possibility appears to be all the more worth consideration as many organism incorporate sizable amounts of magnesium into their hard parts by forming magnesian calcites (see Table 13). This definitely attests to the presence of the cation in the body fluids at least of the groups of organisms concerned. A correlation between skeletal mineralogy and the magnesium concentrations present in the blood of the organisms (or even in such body fluids as secrete the carbonate) does not seem to have been attempted so far. Research in this direction would certainly contribute to the clarification of the problem.

KITANO and HOOD (1965) envisaged the possibility that organic compounds which may occur in the body fluids of carbonate secreting organisms have an influence on the polymorphic crystallization of calcium carbonate. They used the method of CO_2 escape, at room temperature, from solutions of calcium bicarbonate in which various organic compounds were dissolved. In the absence of magnesium, only serine and taurine gave rise to small amounts of aragonite in a bulk of calcite, under conditions which would yield essentially pure calcite in the absence of these organic compounds. Most of the other compounds tested were ineffective in that calcite was obtained just as in their absence. Some of the amino acids studied, e.g. glutamate and glycine, caused vaterite to appear in addition to calcite.

The influence of dissolved organic compounds was more clear-cut in the presence of dissolved magnesium. Under conditions yielding mainly aragonite with small amounts of calcite in the absence of organic additives, additions of malate, pyruvate, citrate, lactate, glycylglycine, succinate (sodium salts throughout), chondroitin sulphate, glycogen and a few other materials were found to suppress aragonite in favor of calcite. It thus appears that these compounds counteract the influence of the magnesium to provoke aragonite.

Two possible mechanisms can be suggested through which these compounds may act. (1) They may sequester the free magnesium ion by way of complex formation so that the surface poisoning of calcitic nuclei by the adsorption of hydrated Mg^{2+} (see p. 114f.) is reduced. In this way, calcite has free rein to grow, almost as in the absence of magnesium, and the higher solubility of aragonite is presumably not exceeded. (2) The compounds restoring the predominance of calcite in the presence of dissolved magnesium may obstruct the nucleation of aragonite or be selectively adsorbed on its surface so that the growth of aragonite is inhibited. This latter mechanism may explain the effect of the citrate, the pyruvate and the malate of sodium. These not only counteract the aragonite-provoking influence of magnesium, but KITANO and KANAMORI (1966) have found that their presence also leads to an enhanced incorporation of magnesium into the resulting calcites. It thus appears that these organic anions are able to suppress aragonite through selective adsorption in nearly the same way in which calgon (sodium polyphosphate) has been postulated to inhibit the crystallization of calcite (RAISTRICK, 1949; BROOKS, CLARK and THURSTON, 1950). When aragonite cannot grow due to adsorbed organic anions, and the supersaturation with respect calcium carbonate is thus not released, calcite will be the only phase able to grow, although at slow rates, due to adsorbed magnesium. It is conceivable that in this situation, there is more time for the magnesium adsorbed on calcite to become dehydrated, and calcites with higher magnesium should be the result, just as in the experiments of KITANO and KANAMORI.

Since at least some influence of organic compounds on the polymorphic crystallization of calcium carbonate has now been demonstrated by the experiments of KITANO and coworkers, a vast field for possible investigations has opened up in that many more organic compounds should be tested. At the same time, however, information concerning the complexing properties of the compounds in question, with respect to calcium and magnesium, should be considered and determined, if necessary. Furthermore, their adsorption on calcite and aragonite must be studied in order to facilitate the decision as to which mechanism is leading to one or the other polymorph with a given organic compound.

In papers discussing the origin of skeletal carbonates, the term "matrix effect" is frequently used (see e.g. WILBUR and WATABE, 1963; and papers in SOGNNAES (editor), 1960). The term is obviously intended to summarize various influences of the surrounding or adjacent organic matter, essentially solid in consistency, on the formation of skeletal carbonate. The most stringent mechanism would be that calcite or aragonite are deposited according to the atomic periodicities of the organic substratum which coincide with those of one of the two carbonates, thus giving rise to epitactic nucleation. However, there seems to exist no evidence for epitactic relationships between carbonates and solid organic substrata (see discussion to WILBUR and WATABE, 1963; and TRAVIS, 1963). Such an influence does not appear very likely, since the interatomic periodicities, if any, of such organic substances as proteins are different in order from the periodicities occurring in the carbonate minerals.

A substance showing a very distinct "matrix effect" is aragonite itself. Aragonitic nuclei once present, whatever their origin, will give rise to the further crystallization of aragonite, even under conditions which would otherwise lead to calcite, as has been mentioned on p. 105 and p. 111.

Thus, the polymorphic crystallization of calcium carbonate is not determined directly by organic substrata, but rather by the chemical composition of the body fluids involved, or possibly by (inherited?) inorganic nuclei. In contrast, the fabrics of carbonates which occur in organic hard parts, and which can be of an amazing variety, e.g. in mollusc shells (Bøggild, 1930; Horowitz and Potter, 1971) and in marine algae (Pobeguin, 1954), can be understood only from the standpoint of the surrounding organic structure. For example, Travis (1963) has shown that the calcite fabrics in decapod crustacea form in close association with histologic structure. It is nevertheless possible that the purely inorganic mechanisms, proposed on pp. 117—131 for the different orientations of aragonite crystallites in oöids may be active to some extent also in organisms and thus contribute to the variety of organic carbonate fabrics.

5. The Persistence and Transformation of Aragonite

On pp. 98 ff. it was shown that any aragonite occurring under earth-surface conditions is unstable with respect to calcite. In the following chapters (C. II. 1 to C. II. 4) various mechanisms leading to the metastable formation of aragonite have been discussed. Of these possibilities, the inorganic precipitation from sea water is known to supply calcareous sediments with appreciable amounts of aragonite. More aragonite is contributed in the form of organic skeletal parts. Ancient limestones, however, consist chiefly of calcite, and in rocks older than about Tertiary, remnants of aragonite occur, but only rarely. The transformation of aragonite to calcite must, therefore, take place at some time during the same interval in which the originally loose calcareous sediment converts to an essentially hard limestone. Before discussing to what extent the two processes may be interrelated it is advisable to examine more closely the conditions under which the conversion aragonite–calcite takes place.

From the experience that, despite its metastability, aragonite occurs as a mineral and may be kept unaltered in mineral collections for indefinite times, it must be concluded that at normal temperature the transformation to calcite is inhibited by considerable kinetic difficulties. As mentioned on p. 97 dry aragonite converts to calcite at perceptible rates only when heated as high as about 400° C. In this range the temperature dependence of the transformation was studied by Chaudron (1952). Reaction rates were found to vary considerably with the mode of formation of the aragonite studied. The resulting activation energies range between 45 and 80 kcal/mole. These results, when extrapolated to temperatures below 100° C (Fyfe and Bischoff, 1965), lead to reaction times of the order of tens of millions of years, thus confirming the prolonged persistence of dry aragonite under normal conditions. DAVIS and ADAMS (1965) calculated an activation energy of 106 kcal/mole from their rate studies of the transformation of a natural aragonite between 380° C and 500° C. Their extrapolation to ordinary temperatures results in reaction times of the order of 10^{40} years. The conclusion is that at earth-surface temperatures dry aragonite should persist for practically unlimited periods of time.

When we realize that the transformation is considerably accelerated by the presence of water (see below), the persistence of dry aragonite seems to have no bearing on the occurrence of the mineral in sediments, considering the ubiquity of water. However, "dry" conditions, i.e. environments essentially free from water, may exist in bituminous and oil-bearing sediments. It is indeed from asphalt and oil shales of Carboniferous age that the oldest fossil shells with preserved aragonite have been reported (STEHLI, 1956; HALLAM and O'HARA, 1962). FÜCHTBAUER and GOLDSCHMIDT (1964) have described well-preserved aragonitic mollusc shells from oil-impregnated rocks of Wealden age (Lower Cretaceous) from NW Germany. Such occurrences are of great interest to paleontology because the original textures of ancient fossil shells may be studied. These are normally obliterated in originally aragonitic organisms by recrystallization to calcite (SORBY, 1879; BØGGILD, 1930; see also HOROWITZ and POTTER, 1971). Another important result is the conclusive demonstration that the transformation aragonite–calcite does not proceed in the absence of water, not even in the course of hundreds of millions of years. This means that a conversion in the dry state is indeed completely inhibited in the temperature range characteristic of diagenesis and is, therefore, of no importance in the evolution of sediments.

Even the restrictions imposed on water circulation by low permeability, e.g. in argillaceous sediments, appear to favor the preservation of aragonite, as evidenced by the occurrence of aragonite-bearing ammonite shells in shales of Jurassic age (FÜCHTBAUER and GOLDSCHMIDT, 1964; ANDALIB, 1970). However, in this and similar cases, special factors [e.g. the persistence of organic sheaths enclosing the aragonite or the presence of sea water (see below)] must have been active which prevented the transformation before the sediment became compacted to such a degree as to become practically impermeable.

The fact that the conversion of aragonite to calcite proceeds in contact with water at 100° C and below is evident already from the experiments published by ROSE (1837). This author found that aragonite precipitated from hot solutions must be filtered and dried immediately in order to avoid substantial admixtures of calcite. Although it was recognized that the transformation is an important factor in the preservation of fossil shells (SORBY, 1879; BØGGILD, 1930) and constitutes one of the major steps in the evolution of limestones, almost all of the systematic knowledge concerning the wet transformation is of very recent date and is based on the experimental research of three authors (BISCHOFF; FYFE; TAFT).

At 100° C, aragonite immersed in pure water converts to calcite in the course of about a day (FYFE and BISCHOFF, 1965). At room temperature the time required for complete transformation is of the order of several months (TAFT, 1967). The increase in reaction time between 100° C and about 60° C is characterized by an activation energy of 57.4 kcal/mole (BISCHOFF, 1969). Although this is of the same order as the values determined by CHAUDRON for the dry reaction near 400° C, the greater rates due to the presence of water can be explained only by a reaction mechanism different from a solid-state reaction. It must be envisaged that the wet reaction proceeds via solution, and the governing aspect must be the greater solubility of aragonite in comparison to calcite (p. 99). Aragonite imparts a certain concentration to the solution and the dissolved ions reprecipitate in the form of less soluble calcite.

The reaction mechanism via solution is illustrated by the changes in reaction rate caused by the presence in solution of various dissolved ions. Of the species related to $CaCO_3$, Ca^{2+} added as calcium chloride has a distinctly accelerating influence. The same is true for dissolved CO_2 gas (BISCHOFF and FYFE, 1968). The latter may be explained by the influence of Ca^{2+} ions as well, since their concentration is increased by dissolved CO_2. The influence of HCO_3^- and CO_3^{2-} has not so far been tested directly by addition of alkali bicarbonate or carbonate, but from the negative influence of alkali hydroxide (BISCHOFF and FYFE, 1968) it must be concluded that CO_3^{2-}, whose concentration is enhanced by OH^- according to (11), retards the conversion. Thus the reaction rate is not influenced in a symmetric fashion by an excess of one of the component ions Ca^{2+} and CO_3^{2-}. This may be related to the difficulty of the CO_3^{2-} ion to become properly oriented when adsorbed in excess on a growing calcite surface (see p. 130 and Fig. 48). However, it must be emphasized that alkali hydroxide has only a retarding influence and that it does not completely inhibit the transformation when present at moderate concentrations, i.e. below the solubility product of $Ca(OH)_2$.

It is not possible here to describe all of the experiments of BISCHOFF and FYFE. On the one hand, these authors, as well as TAFT (1967), tested the influence of quite a number of cosolutes at various concentrations and also in combinations, so that their work can be fully appreciated only by consulting the original papers. On the other hand, many crucial experiments, e.g. with additions of soluble carbonates, remain still to be done. The most important results are summarized below chiefly in a qualitative fashion, because similar to the dry reaction, the absolute rates of the wet transformation as well depend to some extent on the origin of the aragonite used.

Such ions as are not related to $CaCO_3$ nor form sparingly soluble salts with the component ions Ca^{2+} and CO_3^{2-} act as accelerators throughout. BISCHOFF and FYFE (1968) came to the conclusion that the reaction rate increases with the ionic strength of such cosolutes. A similar influence is known to occur in many chemical reactions, and it is not clear so far which step in the transformation aragonite–calcite could be particularly affected by ionic strength.

Those ions which are known to form sparingly soluble salts with one of the component ions of $CaCO_3$, e.g. SO_4^{2-} and bivalent cations, show a retarding or inhibiting influence. According to BISCHOFF and FYFE, Sr^{2+} and Ba^{2+} do not act as inhibitors when applied at concentrations below that at which their carbonates would precipitate.

The most marked inhibitor is the magnesium ion. Its retarding influence is observed even at a molality as low as 0.0001 (BISCHOFF and FYFE). According to experiments by TAFT (1967) at room temperature, a molality of about 0.001 is sufficient to cause a complete suppression of the transformation aragonite–calcite, at least for the duration of several months. A similar effect is observed also at higher magnesium concentrations, notably at about 0.05 M, which corresponds to the magnesium level of sea water. Here no change could be detected after more than a year.

From their detailed kinetic data BISCHOFF and FYFE come to the conclusion that nucleation and growth of calcite are the most important processes which control the aragonite transformation via solution. Thus, the activation energy of 57.4 kcal/mole determined by BISCHOFF (1969) from the reaction rates in pure water between 60° C and 100° C must be composed of contributions from both processes. This is perhaps the reason why the activation energy is distinctly greater than the interaction energy of the calcium ion with the water dipole of 44.9 kcal/mole (Table 20). The latter should be expected as the maximum for the activation energy if the crystal growth of calcite were the only rate-determining process.

It is reasonable to conclude that either nucleation or growth of calcite, or both, are affected by the inhibiting action of the Mg^{2+} ion. Such a conclusion is in harmony with the explanation given in the last paragraphs of chapter C.II.2.b) for the ability of dissolved magnesium to cause the precipitation of $CaCO_3$ as aragonite. This phenomenon was ascribed to the inability of calcitic nuclei to grow when encumbered by adsorbed magnesium ions still hydrated on the outside. The possibility that the dehydration barrier of the magnesium may also be an obstacle to the nucleation of magnesium-bearing rhombohedral carbonates was not excluded. It thus appears that the Mg^{2+} ion acts through the same mechanism both in the formation of aragonite and in its preservation. BISCHOFF (1968) tries to make a case for inhibited nucleation of calcite as prevailing in importance over crystal growth, although he and FYFE did not study the two processes separately. However, as mentioned on p. 111, purposely added calcite nuclei are ineffective when enough magnesium is present in solution to cause the precipitation of aragonite. Experiments on the transformation of aragonite with added calcitic seeds do not seem to have been carried out so far. When the ineffectiveness of calcitic seeds in the formation of aragonite is considered, it is legitimate to infer that also in the inhibition by magnesium of the aragonite transformation, the nucleation of calcite is less important as a kinetic obstacle than crystal growth.

Of the other small cations, Fe^{2+} and Ni^{2+} were tested by FYFE and BISCHOFF (1965). They appear to exert an inhibiting influence similar to that of Mg^{2+}. This behavior, notably of Fe^{2+}, may be the reason not only for the formation of aragonite in certain ore deposits (e.g. in the Erzberg, Austria), but it may also contribute to the preservation of the mineral.

Nevertheless, since it is so abundant in sea water, magnesium is geologically the most important inhibitor, especially in view of the fact that most aragonite occurring on the earth has formed in the sea. From experiments at $108°$ C, BISCHOFF and FYFE (1968) deduce for the SO_4^{2-} ion an inhibiting influence similar to that of Mg^{2+}. However, the experiments with Na_2SO_4, summarized in Tables 23 and 25, show that calcite is quite able to nucleate and to grow at considerable SO_4^{2-} levels. Therefore, the SO_4^{2-} ion may be expected to act merely as a retarding agent, unless its concentration is so high that the aragonite becomes coated with a sparingly soluble calcium sulphate. In normal sea water the retarding influence of SO_4^{2-} may be considered unimportant in comparison to the inhibiting power of Mg^{2+}.

The ability of sea water to preserve aragonite is reflected by the mineralogical composition of young marine carbonate sediments, among which in particular those deposited in shallow water contain aragonite as a main constituent. Published analyses warrant the generalization that aragonite is normally unable to transform to calcite as long as the sediment is in contact with sea water. Of course, this rule applies only when the sea water in question is saturated with aragonite as is the case in warm shallow seas, or when water of marine origin fills the pores of a calcerous sediment, i.e. when an excess of calcium carbonate is present to maintain saturation. Otherwise dissolution will take place. By this process, the aragonite of organic hard parts is, as a rule, dissolved preferentially compared to calcitic remains. The result is a concentration of calcite, which is the characteristic carbonate of the sediments of deeper seas (see e.g. PILKEY and BLACKWELDER, 1968).

It is understandable that most observations concerning the preservation of aragonite in sea water refer to Recent and Pleistocene sediments. However, examples of aragonite preserved by sea water in deposits as old as the Miocene are known as well. Unconsolidated aragonite sediments up to this age, which have soaked in sea water since they were deposited, were encountered in drilling operations in the Pacific atolls of Bikini and Eniwetok at various depths down to a maximum of about 800 meters (SCHLANGER, 1963).

In contrast, much younger carbonate sediments, of Pleistocene age, which have been removed from the marine environment by way of epirogenetic movements and/or sea-level fluctuation, show varying amounts of calcite formed at the expense of aragonite; e.g. in Florida (GINSBURG, 1957; STEHLI and HOWER, 1961), the Bahamas, the Bermudas and elsewhere (FRIEDMAN, 1964; GAVISH and FRIEDMAN, 1969). Obviously the transformation of aragonite to calcite can proceed only after meteoric waters have flushed away the sea water from emerged sediments. When substantial amounts of magnesium are no longer present in the fresh water, which now fills the pore space, at least partly or occasionally, existing calcitic nuclei are no longer encumbered by sheaths of tightly hydrated magnesium ions and have free rein to grow at the expense of the more soluble aragonite. The same applies to any new calcitic nuclei which may now form.

The complete calcitization of an aragonitic sediment under the influence of fresh water may take about 100 thousand years or more (MATTHEWS, 1968; GAVISH and FRIEDMAN). In detail, the degree of transformation in coeval sediments depends upon the availability of interstitial (fresh) water. For example, the Pleistocene Miami Oölite of Florida appears to contain considerably more calcite below the ground-water table than above, where interstitial water is not continuously present (GINSBURG, 1957).

In the Pleistocene carbonate sediments which are exposed on the island of Barbados, more aragonite is preserved in regions with low rainfall, whereas comparable sediments in regions with high rainfall show a fairly complete transformation to calcite (MATTHEWS, 1968).

In all occurrences of emerged Pleistocene carbonates showing progressive transformation to calcite, the reaction is accompanied by increasing lithification, tending to result in solid limestones. Lithified zones consisting of calcite were encountered in the drilled sections of Tertiary carbonates at Eniwetok and Bikini, which are otherwise remarkable for the occurrence of unconsolidated aragonite sediments, as mentioned above. Occurrences of terrestrial fossils indicate that the calcitic zones mark episodes of emergence above the level of the sea. It is thus legitimate to ascribe the calcitization of the originally aragonitic sediment and the concomitant solidification to the influence of meteoric water also in these Tertiary sections (see SCHLANGER, 1963).

The replacement of sea water by interstitial fresh water thus appears to be essential for the solidification of marine calcareous sediments. Although the latter process is merely a corollary to calcitization, it is through the action of both calcitization and solidification that young carbonates may become similar to limestones (as we know them from ancient formations), or at least approach them in character. The widespread operation of the mechanism in younger sediments strongly suggests that it is perhaps of more general importance and that it has played a decisive part also in the diagenesis of ancient limestones. The presence of fresh water may be regarded as the general prerequisite for triggering the transformation of loose aragonitic sediments to hard calcitic limestones. In this connection it is interesting to note that limestones deposited in fresh-water environments (calcareous tufa, travertines) are in most cases calcitic and lithified from the beginning (see e.g. IRION and MÜLLER, 1968; SAVELLI and WEDEPOHL, 1969).

Recently discovered occurrences of submarine lithification (FISCHER and GARRISON, 1967; MACINTYRE, MOUNTJOY and D'ANGLEJEAN, 1968; MILLIMAN, ROSS and TEH-LUNG KU, 1969; SHINN, 1969; LAND and GOREAU, 1970) seem to be at variance with the view that fresh water is the main agent in limestone diagenesis. However, in all known cases where carbonates were lithified in contact with sea water, the cement was found to consist of aragonite and/or magnesian calcites. Essentially pure calcite ("low-Mg calcite") has not been found so far. Evidently submarine lithification does not proceed in the direction of the constitution of ancient limestones. The occurrence of aragonite cement is in harmony with the conditions of formation of the mineral as described on pp. 107—113. In shallow seas, CO_2 loss either through the surface of the sea or by the photosynthetic activity of plants, may produce either cementation by aragonite or lead to the precipitation of loose particles, depending upon the hydrodynamic conditions (cf. BATHURST, 1968). In the intertidal zone, in the lithification of beachrocks, in which aragonite is the most common cement, evaporation is suggested as an additional factor (GINSBURG, 1953; STODDART and CANN, 1965).

The conditions for the precipitation of magnesian-calcite cements from sea water are not well understood. It is conceivable that magnesian calcites form very slowly when the supersaturation of the sea water with respect to $CaCO_3$ is not high enough for aragonite to precipitate. At least in some instances, e.g. in the case of the lithification described by MILLIMAN et al. in sediment cores from the Red Sea, the layers cemented by magnesian calcite are found below the surface of the sediment. In most other instances, it cannot be ruled out that magnesian calcites formed under a cover of sediment now eroded. In a buried sediment the interstitial sea water may become enriched in dissolved CO_2 from decaying organic matter. In this manner, the solution capacity for $CaCO_3$ is enhanced, and by actual dissolution of $CaCO_3$ from the sediment, the ratio Mg^{2+}/Ca^{2+} may become lowered to such an extent as to allow reprecipitation in the form of magnesian calcite.

Furthermore, decaying organic matter may affect the composition of the interstitial fluid in a variety of ways. For example, in conjunction with the bacterial reduction of sulphate, carbonate ions may be produced [see Eqs.(30) and (31), p. 182] which lead to the precipitation of aragonite in the pore space. This may be the chief mechanism giving rise to cementation by aragonite below the surface of a sediment. The process implies that the cemented layers or patches were originally higher in organic matter than the adjacent unconsolidated sediment.

Interstitial precipitation of carbonate due to bacterial sulphate reduction may occur also in non-carbonate sediments. For example, it may be considered as an explanation for the origin of calcareous (septarian) concretions occurring in argillaceous sediments. The low values for the heavy carbon isotope ^{13}C encountered in such carbonate nodules indicate that their formation must be connected, in some way or other, with concentrations of organic matter (HOEFS, 1970). Up to now, the most discussed process presumed to lead to the formation of carbonate concretions has been the decomposition of proteins (see e.g. LIPPMANN, 1955; HOEFS). The ensuing liberation of ammonia may give rise to the precipitation of some carbonate cement also in calcareous sediments, notably in layers rich in organic remains. However, the quantity of nitrogen contained on the average in proteins and the amounts of ammonia to be expected from their decay are too low to afford a satisfactory explanation for the amounts of carbonate contained in the concretions from argillaceous environments. Sulphate reduction is capable of supplying considerably greater amounts of carbonate ions, in that it proceeds from organic carbon. It is thus not limited to proteins but is known to act upon many kinds of organic matter.

The foregoing digression on submarine cementation has shown that the cements formed from interstitial solutions related to sea water do not lead to rocks of essentially calcitic composition as we are accustomed to finding them in ancient formations. The occurrences of submarine lithification (cementation) reported so far in the literature do not warrant an appraisal of the volumetric importance of the phenomenon. However, except perhaps in the case of (coral) reefs, the process does not appear to affect greater thicknesses of sediment. At least in the occurrences studied by MILLIMAN et al. (Red Sea; see also GEVIRTZ and FRIEDMAN, 1966) and SHINN (Persian Gulf), cemented layers alternate with, or are embedded in, unconsolidated sediment. From this point of view, submarine cementation, if stratiform in character, may be an important factor in the preservation of (undeformed) fossils and sedimentary structures, such as bedding planes including the markings of them. Penecontemporary erosion of cemented layers may give rise to new structures, such as intrastratal conglomerates.

However, submarine cementation cannot be final in character, since the metastable cements involved will start converting to calcite as soon as they come into contact with water free from magnesium. It is known that in the course of the conversion of aragonite, new pore space usually develops by the leaching of certain types of aragonitic particles, such as shells (BATHURST, 1966) or oöids (FRIEDMAN, 1964, 1968; "moldic porosity"); and perhaps more important, aragonitic domains tend to become friable (see e.g. ROBINSON, 1967) when part of the mineral is being leached and subsequently redeposited as a calcite cement elsewhere, e.g. in primary pores. Notably for this and similar reasons submarine cementation cannot be accepted as a step on the way toward ultimate consolidation.

Therefore, irrespective of the possibility of cementation in the marine environment, water free from magnesium appears to be indispensable as an agent in the diagenesis of limestones. In this connection it is interesting to look for possible mechanisms other than the migration of meteoric water by which magnesium may be removed from a marine brine filling the pore space of a sediment. Although it is hard at present to assess the efficiency of such mechanisms, most of them not being well understood, the following possibilities may be considered in a tentative fashion: dolomitization, the formation of chlorites in adjacent argillaceous beds, and ionic filtration through such beds. However, in view of what is known concerning the trends of strontium in calcareous sediments, this element must then be removed as well along with magnesium, but this would be at variance with the general geochemical behavior of strontium in the three processes considered. The conclusion is thus inevitable that virtually all ancient limestones have undergone diagenesis in fresh water of ultimately meteoric origin at some time during the course of their history.

The situation is illustrated by the following sketch of the geochemistry of strontium in calcareous sediments. It is based chiefly on KINSMAN (1969). FLÜGEL and WEDEPOHL (1967) and BAUSCH (1968) have reasoned along similar lines. Aragonite forming from sea water by inorganic precipitation, e.g. in the form of oöids, contains about 1% strontium. This is in fair agreement with the ratio Sr^{2+}/Ca^{2+} in sea water, which is about

10^{-2}, and with the distribution coefficient with regard to strontium,

$$\frac{[Sr^{2+}_{cryst}]/[Ca^{2+}_{cryst}]}{[Sr^{2+}_{solut}]/[Ca^{2+}_{solut}]}$$

which is about unity for the precipitation of aragonite, according to the experimental determinations by HOLLAND and coworkers (see KINS-MAN, 1969). The aragonitic hard parts of most marine organisms contain about 1% strontium, or generally slightly less. Aragonitic molluscs show considerably lower concentrations, around 0.2%. The deviations from the relative concentration in sea water are presumably conditioned by selective incorporation of inorganic cations into the body fluids and/or specific peculiarities of organic carbonate precipitation. Both these processes are far from understood so far. The strontium contents of recent marine aragonitic sediments are also about 1%, or less, depending on the chief constituents, in particular on the amount of calcitic skeletal parts present. As a rule, these contain on the order of 0.1% strontium, in keeping with the distribution coefficient for the precipitation of calcite, for which a value of 0.14 has been determined at $25°$ C in the pure system.

If an aragonitic sediment is calcitized in a nearly closed system, i.e. following the exchange of the marine brine by fresh water in one step, the original strontium content would be essentially retained in the ensuing limestone. However, the concentrations in ancient limestones are in general of the order of 0.0X%. The decrease from the value in the original sediment of about 1% to 0.0X% can be explained only by assuming that extremely large quantities of meteoric water have passed through the pores of ancient limestones during their diagenetic history. According to an estimate given by KINSMAN (1969) on the basis of the above quoted distribution coefficients, the quantity of fresh water required is in excess of 10^5 pore volumes.

Water quantities of this order may appear excessive at first sight, and a lower estimate would indeed result if it should turn out that the experimental distribution coefficient of 0.14 is high. Such a trend is very likely when the rather short durations of most laboratory experiments are considered. Since strontium is less firmly hydrated than calcium (Table 21) it must be expected to be incorporated into calcite in excess over the equilibrium value when strontium is rapidly coprecipitated at ordinary temperature. Nevertheless, water quantities of 10^5 pore volumes are commensurate in order of magnitude with the time of about 100 thousand years characteristic for the calcitization of emerged young carbonate sediments. However, longer times must be envisaged perhaps for carbonate formations of greater thickness, say of hundreds of meters.

Even so, irrespective of the exact quantities of fresh water required for calcitization and for the removal of strontium from a marine carbonate sediment, it is evident that fluid transport through the pore space must play an eminent rôle in the diagenesis of limestones. When aragonite transforms in a flowing interstitial solution, the mineral may be dissolved in one place, e.g. where undersaturated water enters into a carbonate formation, and reprecipitation of calcite may occur at a different place along the path of the interstitial fluid. In this manner, the pore space at some depth may become filled by calcite at the expense of aragonite leached out elsewhere. Various models depicting simple characteristic cases were discussed by MATTHEWS (1968). On a smaller scale, domains differing in primary grain size and permeability (e.g. the corals in the Pleistocene of Barbados; PINGITORE, 1970) may react at different times and yield calcite fabrics characterized by differences in grain size. In this situation, in spite of complete recrystallization, certain critical flow rates may lead to the preservation of originally aragonitic textures, such as fossils and oöids, which would otherwise be obliterated.

However, it does not seem appropriate to draw too many conclusions from emerged young formations and try to apply them in great detail to the interpretation of stratiform ancient formations. In these, the ground-water flow may be expected to have followed essentially the most permeable beds (see v. ENGELHARDT, 1967). A great variety of settings and ensuing flow patterns are imaginable in bedded sequences, the discussion of which would be beyond the scope of the present monograph.

In general, reliable information concerning the flow pattern of the pore fluid during diagenesis is hard to obtain, although intelligent guesses are sometimes possible for simple geologic structures. The time when fresh water may have started entering a calcareous marine formation can occasionally be determined from overlying lacustrine sediments (LIPPMANN and SCHLENKER, 1970).

It is safe to assume that the diagenesis of extensive stratiform carbonates is characterized by flow rates which are much lower than those to which outcropping Pleistocene carbonates are now exposed. Estimates of millions of years for the duration of calcitization and solidification do thus not appear excessive. Consequently, the influence of overburden and lithostatic pressure must be taken into account, particularly, in thick continuously marine sequences.

A sediment in which aragonite transforms to calcite via solution must be expected to yield to pressure by small-scale differential movements of the order of the grain size. As the transformation proceeds the fabric will collapse where aragonite disappears. Calcite will deposit preferably in those preexisting and newly forming pores which by virtue of their small size are most likely to resist the lithostatic pressure. This type of readjustment will continue until the transformation to calcite is complete. It can be visualized that primary aragonitic fabrics may thus become obliterated and that the resulting calcite is of much finer grain size than the original sediment. A great many fine-grained limestones described as "micrites" and explained as "calcilutites" may have originated by this mechanism, which can be referred to as "pressure-conditioned transformational comminution". The differential movements, which normally leave no trace within the beds they affect, are often attested to by stylolite seams which have developed from originally smooth bedding planes.

The most important consequence of lithostatic pressure is, of course, compaction. The process starts as soon as more sediment accumulates on top of a given layer, and porosity usually decreases steadily with increasing overburden. In the beginning, rearrangement of particles is the predominant mechanism and with greater overburden, pressure solution becomes active (v. ENGELHARDT, 1960). This is the general behavior of loose sediments, among them carbonates. Carbonate sediments which have undergone (submarine) cementation will at first resist compaction until lithostatic pressure is great enough for pressure solution to become appreciable.

In addition to the more or less continuous compaction under increasing overburden, both loose and cemented carbonate sediments will show a discontinuous decrease in porosity when the aragonite they contain transforms to calcite. Porosity will then be lost not only because the transforming sediment is particularly prone to yield to lithostatic pressure, but also because the resulting calcite, in vigor of its smaller density (2.71), occupies a greater volume than the original aragonite (2.93). An increase in solid volume of about 8.1% is calculated from the densities. A pore-space reduction up to 7.5% (of the total final volume) may thus be accounted for by the conversion of aragonite. In most cases the reduction in porosity due to the conversion is smaller than 7.5%, depending on the original percentage of aragonite and the final porosity.

Values below about 5% porosity seem to be characteristic of dense limestones from ancient formations, e.g. from the German Jurassic (HE-LING, 1968). In these rocks, most of the porosity was found to consist of closed (unconnected) pores. The open porosity, which includes pore sizes as small as 100 Å, was found to be in the order of 0.5%. Except for the fractures they may contain, such dense limestones are rather impermeable and will resist further reactions involving solutions and transport of same. It is hard to conceive that such rocks are still susceptible to complete dolomitization. The leaching of the strontium which the sediments may have contained in their aragonitic phase must be complete before the pore space is reduced to essentially closed porosity. In the case of diagenesis under overburden, this stage will probably be reached shortly after the last portion of aragonite has converted to calcite. The complete reduction to zero porosity is very likely a slow process which extends over extremely long periods of time, commensurate perhaps with the duration of geologic eras.

The outline given in the last paragraphs on the diagenesis of marine carbonate sediments under overburden is certainly highly simplified and speculative, chiefly because the process is not susceptible to direct observation. The main emphasis has been on the transformation aragonite–calcite, and alternative models involving intermediate stages were left out of account. For example, under the flow regimes characteristic of buried sediments, the complete removal of magnesium from the pore solutions is a slow process. A stage may be conceived where magnesium is low enough for calcite to form but still sufficiently high for sizable amounts of magnesium to be incorporated into calcite. Magnesian calcites, notably those of inorganic origin, are known to occur in very small crystal sizes. (This is most likely due to the strain imposed on the crystal structure by the disordered distribution of the magnesium ions and correlates with the poor stability of magnesian calcites; see p. 161 ff.). On the other hand, the crystal orientation of magnesian calcites is known to persist in the essentially pure calcites which result from the leaching of Mg and ensuing recrystallization. The best known example is the preservation of crystal orientation in crinoid stems and other remains of cchinoderms. This type of conservation may be attributed to epitactic nucleation of the final calcite on the preexisting magnesian calcite. A similar relationship may be conceived to exist between the very fine grain size of dense limestones ("micrites") and fine grained magnesian calcites of some intermediate stage of fresh-water diagenesis. This would be an alternative mechanism, different from "pressure-conditioned transformational comminution", leading from coarse-grained aragonitic sediments to dense calcitic limestones.

The problem of the preservation of sediment structures and fossils, in spite of the complete transformation via solution of original aragonite, has been touched upon several times. Differences in permeability and penecontemporary cementation were mentioned from among the favorable factors. The preservation of such structures is reminiscent of pseudomorph formation among minerals. This process may be studied by experiments, and crystal shapes are found to be preserved despite radical changes in chemical composition when the original crystal is treated with appropriate solutions. Examples are the conversion of halite to AgCl by $AgNO_3$ solution (CORRENS, p. 309) and of calcite to fluorite (CaF_2) by hydrofluoric acid (GLOVER and SIPPEL, 1962). The latter authors quote examples from the paleontological literature (e.g. UPSHAW et al., 1957) where the calcium carbonate of microfossils was converted to fluorite by immersion in hydrofluoric acid. In these cases details of the shell structure were found preserved as fluorite in excellent condition.

The more or less perfect preservation of originally aragonitic fossils (and oöids), despite complete transformation via solution to calcite may also be understood as a kind of pseudomorph formation. CORRENS (p. 309) shows that morphological preservation is favored by high supersaturation. Differences in preservation of originally aragonitic structures should be studied from this viewpoint. Although many more parameters will have to be considered than in the simple reaction

$$NaCl + Ag^+ \rightarrow AgCl + Na^+$$

studied by CORRENS, such observations, supplemented perhaps by experiments, might be a substitute for more direct investigations in natural environments, i.e. in buried sediments, of the processes which lead to the preservation of aragonitic structures.

D. The System $CaCO_3$–$MgCO_3$

I. The Dolomite Question

Among the constituents of ancient carbonate rocks, dolomite, $CaMg(CO_3)_2$ is second in importance after calcite. Although both minerals may be found associated in dolomitic limestones, most dolomite occurs in essentially monomineralic bodies. These range in dimension from patches or layers of the order of centimeters to formations of considerable thickness. In most cases, the small-scale bodies occur embedded in limestone as a host rock (see e.g. LIPPMANN and SCHLENKER, 1970), sometimes also in argillaceous rocks, e.g. in the European Keuper. Moreover, dolomite may be present as a minor constituent in almost any sedimentary rock.

The term dolomite has been used both as a rock name and for the mineral. The problem of the genesis of the rock type and the mineral has intrigued earth scientists from the beginnings of geology, and has given rise to a vast number of publications. These have been summarized from time to time in monographs dealing with the dolomite question, e.g. by VAN TUYL (1916), KROTOV (1925), FAIRBRIDGE (1957), STRAKHOV (1956, 1958) and USDOWSKI (1967).

The older publications are of limited interest today, except for the occurrences they describe. This is true not only of the many speculations concerning the genesis of dolomite. In addition most of the published experiments claiming successful syntheses of dolomite must be regarded with extreme caution, unless the final product was identified by X-ray diffraction and confirmed by the presence of superstructure reflections (see p. 30).

The formation of dolomite is a rather clear-cut problem when the structural definition of the mineral is considered. Complications arise, however, from the occurrence of varieties deviating in cation disorder and excess of calcium, as reviewed on pp. 47—50 (also from the existence of huntite, $CaMg_3(CO_3)_4$, a second stoichiometric phase in addition to dolomite intermediate between calcite and magnesite).

The problem of the origin of dolomite arose and persists today for two main reasons: (1) In spite of the simple composition, $CaMg(CO_3)_2$, it has not been possible to synthesize the mineral from appropriate solutions in the range from ordinary temperatures to over 100° C, i.e. at temperatures characteristic of the sedimentary cycle (see pp. 107—115).

(2) Dolomite is rare in Recent and Pleistocene sediments in proportion to the total amount of carbonate sediment of that age, and also in comparison with the abundance of dolomite in ancient sedimentary rocks.

For a long time sub-recent sediments were considered devoid of dolomite, and many hypotheses on the origin of the mineral were based on this view. The long periods of time that normally are characteristic of many geologic processes were usually deemed necessary for the formation of dolomite. Recently, the picture has been changed by discoveries of dolomite in sub-recent carbonate sediments from lagoonal and tidal environments. The most important localities are in South Australia (ALDERMAN and SKINNER, 1957; VON DER BORCH, 1965; ALDERMAN, 1965), Florida, the Bahamas (SHINN, GINSBURG and LLOYD, 1965), Bonaire (Netherland's Antilles); DEFFEYES, LUCIA and WEYL, 1965) and the Persian Gulf (ILLING, WELLS and TAYLOR, 1965).

In these ocurrences, the dolomites have a grain size of a few microns and have been identified by X-ray powder diffraction. Superstructure reflections are weak in intensity or absent, and shifts of the general reflections are indicative of magnesium-deficient dolomites. In addition to describing occurrences of sub-recent dolomite, the above quoted authors try to interpret its formation in terms of environment. There is agreement that sufficient supply of magnesium, enhanced by evaporation, must be an important factor. Yet no cogent conditions for the formation of dolomite have been derived from the mode of occurrence of the sub-recent dolomites. If supply and concentration of magnesium were the only factors, then dolomite formation should be more common in sub-recent carbonate sediments than it is actually. From this point of view, the sub-recent dolomites are of little help in explaining the genesis of ancient formations. In particular, it must be considered that the sub-recent occurrences are much more limited, both in lateral extension and thickness, than many ancient dolomite formations. Moreover, the complete cation order often found in the latter and the coarse grain size of the saccharoidal varieties have no equivalent among the sub-recent dolomites.

What can be deduced from the mode of occurrence of sub-recent dolomites is a confirmation of the old hypothesis that dolomite does not precipitate or crystallize directly from sea water or related brines, but forms by reaction of solutions derived from sea water with previously deposited calcium carbonates. Such a mechanism had been inferred by geologists from field observations on dolomite and limestone in ancient formations. The main criteria were the wedging out of dolomite layers in a host rock of limestone and, more generally speaking, the observation that dolomite bodies are often delineated by surfaces which are not strictly, but only approximately, parallel (i.e. subparallel) to primary stratifications (see e.g. VAN TUYL). KROTOV (1925) appears to have been the first to write explicit chemical reactions for the dolomitization of calcareous sediments by magnesium-bearing solutions:

$$2\,CaCO_3 + MgSO_4 \rightarrow CaMg(CO_3)_2 + CaSO_4 \qquad (13)$$
$$\text{(Haidinger reaction)}$$

$$2\,CaCO_3 + MgCl_2 \rightarrow CaMg(CO_3)_2 + CaCl_2 \qquad (14)$$
$$\text{(Marignac reaction)}\,.$$

KROTOV also named the reactions in honor of the savants who first envisaged the respective mechanisms in a qualitative fashion.

The main difference between the two reactions is supposed to be due to the low solubility of calcium sulphate (gypsum, anhydrite) in comparison to calcium chloride. From this and from the frequent association of dolomite with calcium sulphate minerals, it was concluded that the Haidinger reaction should be more efficient as a mechanism of dolomitization than the Marignac reaction. However, sea water and most other natural brines reputed to bring about dolomitization contain both sulphate and chloride. Since it is impossible in ionized solutions to assign a given anion to a particular cation, it has become modern usage to generalize the two preceding reactions as:

$$2\,CaCO_{3\,(solid)} + Mg^{2+} \rightarrow CaMg(CO_3)_{2\,(solid)} + Ca^{2+}\,. \qquad (15)$$

Also this reaction is to be regarded only as an overall reaction intended to describe the total material turnover in dolomitization and not the true reaction mechanism. Since dolomite forms crystals which are essentially ionic in character (see p. 26ff.), the only conceivable mechanism leading from ionized dissolved species to an ionic crystal can be:

$$Ca^{2+} + Mg^{2+} + 2\,CO_3^{2-} \rightarrow CaMg(CO_3)_{2\,(solid)}\,. \qquad (16)$$

On pp. 107—115 it was shown, however, that attempts to precipitate dolomite according to this equation do not yield the desired product in conventional laboratory experiments, not even during extended periods of time of weeks and months which are characteristic of the method of precipitation from homogeneous solutions.

There is no discrepancy between reactions (15) and (16) when it is borne in mind that (solid) calcium carbonate, the starting material in reaction (15), must first dissolve to form the ionic species Ca^{2+} and CO_3^{2-}, before it can participate in a reaction with dissolved Mg^{2+}. The only difference between Eqs. (15) and (16) is that the former imposes certain restrictions on the composition of the reacting solution. When writing Eq. (16) we make no assumption whatsoever as to which relative concentrations may lead to the formation of dolomite. In contrast, Eq. (15) is supposed to imply rather low (starting) concentrations of Ca^{2+} and CO_3^{2-} (by virtue of the sparing solubility of $CaCO_3$) and to be compatible with high Mg^{2+} concentrations; i.e. the importance of a high Mg/Ca ratio in solution is emphasized.

The line of reasoning must be interrupted at this point because the conclusions thus obtained are too general in character. Moreover, artificial dolomite has never been synthesized below about 100° C, neither according to (15) nor by direct coprecipitations according to (16). In order to arrive at more detailed information concerning the possible conditions of formation of dolomite, the available physicochemical data on the system $CaCO_3$–$MgCO_3$ have to be reviewed.

II. Phase Relations in the Dry System

The stable phases of the system $CaCO_3$–$MgCO_3$ equilibrate at reasonable rates only at elevated temperatures. Below about 500° C reactions in the dry system are quite sluggish in the absence of mineralizers, such as water, so that recourse must be made to hydrothermal methods (see e.g. GRAF and GOLDSMITH, 1956). Even above 500° C, it is often useful to add small amounts of Li_2CO_3 in order to promote reactivity. At higher temperatures it is necessary to apply confining pressures to prevent the decomposition of the carbonates involved according to:

$$XCO_3 \rightarrow XO + CO_2 . \tag{17}$$

The phase diagram of GOLDSMITH and HEARD (1961) shown in Fig. 49 is based on such experiments. In certain details which they discuss the data of these authors differ slightly from the results of previous workers, but these differences do not affect the general features of the diagram. Also the necessary increase in confining pressure from 2 to 15 kilobars, may reasonably be expected to yield a polybaric diagram not too different in character from an isobaric diagram valid for constant pressure (see however GOLDSMITH and NEWTON).

At and above 1085° C there exists a complete series of solid solutions from 50 mol% (ideal dolomite composition) to 100 mol% $CaCO_3$. The solid solutions with less than 42.5 mol% $MgCO_3$ are disordered magnesian calcites. Cation order becomes noticeable with more $MgCO_3$, as the ideal dolomite composition is approached. For this composition, cation order decreases with increasing temperature and disappears almost completely above 1200° C. Complete order of the cations prevails from about

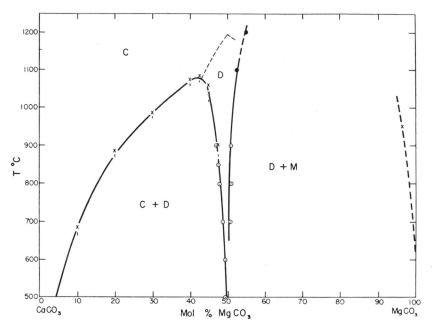

Fig. 49. Polybaric phase diagram of the system CaCO$_3$–MgCO$_3$ from GOLDSMITH
and HEARD (1961). C = calcite (more or less magnesian), D = dolomite, M = magnesite, vertical bars denote two-phase assemblages, crosses mean single phases;
for the experimental points marked by circles, the composition was determined
from X-ray spacings

950° C downwards where the superstructure reflections (10.1), (01.5), and
(02.1) appear with their normal intensities.

From 1075° C (42.5 mol% MgCO$_3$) a two-phase region (miscibility
gap) extends downwards to lower temperatures. The coexisting phases
are magnesian calcites and ordered dolomite of near ideal composition.
Since the top of the two-phase field (42.5 mol% MgCO$_3$) is rather close
to the composition of dolomite (50 mol%), the limb determining the
compositions of the ordered phases is steeper than the one describing the
composition limits of the magnesian calcites as a function of temperature.

The diagram Fig. 49 shows that ordered dolomite can be synthesized
at sufficiently elevated temperatures (and pressures). The same is true for
magnesian calcites of any composition. However, since they would de-
compose on cooling to compositions approaching pure calcite and dolomite, they have to be quenched in order to be preserved for study at
room-temperature.

At first sight, the high-temperature phase diagram, $CaCO_3$–$MgCO_3$, seems to have no relation to the problem of the low-temperature formation in sediments of the phases involved. Yet on the basis of experience with other phase diagrams of similar type involving solids (notably metals), it appears permissible to extrapolate the phase diagram Fig. 49 below 500° C. To some extent, such an extrapolation receives support from results obtained by GRAF and GOLDSMITH (1956) in hydrothermal experiments on the system below 500° C to about 300° C. From this temperature downward equilibration becomes more and more sluggish even in the presence of water.

The extrapolation of the two-phase field(s) of Fig. 49 to earth-surface temperatures leads to the conclusion that essentially pure calcite and ordered dolomite (also magnesite) are the only stable phases in the system $CaCO_3$–$MgCO_3$ under the conditions of the sedimentary cycle. (The metastability of aragonite and hydrous carbonates has been demonstrated in previous sections). All magnesian calcites occurring in the hard parts of organisms and in sediments are thus metastable phases. The same is true for magnesium-deficient and/or disordered dolomites. For example, a magnesian calcite with 20 mol% $MgCO_3$, which frequently occurs in calcareous algae, is stable only above 900° C. At ordinary temperature, the stable association should consist of 40% dolomite plus 60% calcite as discrete phases.

A stability field for magnesian calcites at lower temperatures could be conceived only if the two-phase field would not continue to broaden below 500° C, as assumed in the extrapolation, but would become narrower again towards lower temperatures to form a closed region of immiscibility with a minimum critical temperature. However, this type of a two-phase field is not known to occur in solid systems. The possibility of a closed two-phase field may be discarded as an abnormal case occurring only for liquid immiscibility in systems with water and certain organic compounds, such as nicotine and 2,4-lutidine (see e.g. ANDON and COX, 1952; KORTÜM and HAUG, 1956).

The conclusion is that magnesian calcites, as well as disordered magnesium-deficient dolomites, form at earth-surface temperatures in the stability field of dolomite under conditions where the crystallization of ordered dolomite is inhibited by kinetic obstacles.

III. Systems Involving Aqueous Solutions
1. Hydrothermal Syntheses of Dolomite

If dolomite would readily precipitate or crystallize from aqueous solutions we would know exactly its conditions of formation and there would be no dolomite problem. However, reaction rates are slow up to over 200° C. Nevertheless, given enough time, dolomite can be obtained in hydrothermal runs around 200° C (GRAF and GOLDSMITH, 1956). By the use of precipitation methods adapted to hydrothermal experimentation, the yield can be considerably improved, so that the hydrothermal synthesis of ordered dolomite is no longer a problem (BARON, 1958: 150° C; MEDLIN, 1959: 200° C). In the light of such experiments, it is quite possible that SPANGENBERG (1913) was one of the first to succeed in the hydrothermal synthesis of dolomite, although X-ray diffraction as a tool for its reliable identification was not yet available at that time. SPANGENBERG reacted vaterite, magnesium chloride, and sodium bicarbonate in an autoclave at 180—200° C and 50 atm. He appears to have followed MARC and ŠIMEK (1913) who obtained magnesite under analogous conditions.

USDOWSKI (1967) worked in the range from 180° to 120° C on hydrothermal carbonate systems to which he adapted the methods introduced by VAN'T HOFF for the study of equilibria among evaporite minerals. USDOWSKI determined and discussed a number of phase diagrams involving solutions saturated with calcite, dolomite, and magnesite, as well as with the corresponding chlorides and sulphates. It is not possible here to review USDOWSKI's work in detail. Either his original monograph or his summary in English (USDOWSKI, 1968) must be consulted. One of his most important results is the synthesis of dolomite according to the Marignac reaction (14) at 120° C. At lower temperatures (100° C) the synthesis was no longer successful. Attempts with the Haidinger reaction (13) in the sulphate system were unsuccessful at 120° C and above.

The failure of dolomite to form in the sulphate system and below 120° C in the chloride system did not lead USDOWSKI to the conclusion that dolomite is unable to form under these conditions in nature. He assumed that given enough time dolomite will crystallize also at lower temperatures from the systems he studied, and he used extrapolated versions of his experimental diagrams for the calculation of geochemical balance sheets relating to the quantitative requirements of dolomite formation at lower temperatures.

2. Arguments Against the Hydrothermal Origin of Many Dolomite Formations

The results obtained from the hydrothermal systems might also lead one to the generalization that most natural dolomites have originated at temperatures above 100° C as have been required in the laboratory syntheses so far reported. Such a view would imply that dolomite would form preferably under an overburden of several thousand meters, when normal geothermal gradients are considered. At such depths the supply of the magnesium needed for dolomitization becomes a serious problem, as does the removal of the calcium set free by reaction (15). Even if the carbonate sediments involved would have preserved their permeability despite the lithostatic pressure ensuing from the postulated overburden, impermeable beds must be expected to be present in the overlying strata. Moreover, the environment of sedimentation may change several times in a column several thousand meters thick. A direct exchange of fluids with an overlying sea is thus highly unlikely, if we consider the sea as the most appropriate reservoir for the large quantities of magnesium needed for dolomitization. The brines supplying the magnesium could travel only along permeable pathways essentially parallel to the bedding or via faults. Considerable difficulties will be encountered when the genesis of extensive stratiform dolomite formations is to be explained by hydro-thermal dolomitization. It is indeed conceivable that magnesium-bearing brines may travel long distances along permeable beds, but the most permeable beds will also be the first to become affected by dolomitiza-tion and to lose their permeability under the combined influence of this reaction and the overburden. Dolomitization would also encroach upon adjacent beds whose permeability is less than that of the main pathways, but still sufficient for fluid exchange with the latter and for dolomitiza-tion to take place. In this manner, dolomitic beds grading imperceptibly into limestone should be expected to form.

The knife-sharp contacts which often separate bodies of dolomite from essentially pure limestone are thus left unexplained by the hypo-thesis of deep-seated hydrothermal dolomitization. Also certain small-scale sedimentary features marked by sharp boundaries of dolomite against limestone cannot be reconciled with this mechanism. In addition, the low viscosity of aqueous solutions above 100° C should not favor the formation of interbedded sets of limestone and dolomite layers on the scale of centimeters or the preservation of burrows as dolomite in a host rock of limestone.

In rocks of the Upper Muschelkalk of SW Germany, it could even be shown that, contrary to the setting required by the hydrothermal hy-pothesis, the pore space of limestone layers remained open for a longer time than that of associated dolomite beds. This conclusion of LIPP-MANN and SCHLENKER (1970) is based on the abundance of authigenic quartz and feldspar, which was found to be greater in the limestone layers than in the interbedded dolomite. It must, of course, be postulated that open pore space and ensuing short-range permeability are essential prerequisites for mineral transformations to proceed in the insoluble residue of a carbonate sediment.

Finally, there are a great many dolomite formations which were demonstrably never buried deeper than a few hundred meters. In view of the failure of the hydrothermal hypothesis to account for a number of observations relating to sedimentary dolomite rocks, it is reasonable to continue the search for mechanisms which can explain dolomitization at lower temperatures.

3. The Solubility of Dolomite at 25° C

In the preceding chapters, the question whether or not dolomite is stable at earth-surface temperatures has been discussed on the basis of extrapolations from high-temperature data and field evidence. The question may be approached more directly from the solubility of dolomite. As in the case of the other sparingly soluble carbonates, calcite, aragonite, and magnesite, the solubility of dolomite is governed by such variables as CO_2 pressure, alkalinity, and pH, which are interrelated by the equilibria (9), (10), and (11)(p. 103). The solubility itself may be expressed by the solubility product:

$$[Ca^{2+}] \cdot [Mg^{2+}] \cdot [CO_3^{2-}]^2 = k_{sD} \qquad (18)$$

the type of which follows from Eq. (16).

For the purpose of comparing the solubility of dolomite with that of the simple carbonates, such as calcite and magnesite, we may define an average solubility product:

$$k_{av} = [X^{2+}] \cdot [CO_3^{2-}] = \sqrt{k_{sD}}. \qquad (18a)$$

Since dolomite cannot be synthesized at 25° C, its solubility can be determined only by dissolution experiments. The easiest way to control the partial pressure of CO_2 is to bubble the gas through the solution, i.e. to work at approximately one atmosphere CO_2 pressure.

The first determination of the solubility product (18) by this method was carried out by HALLA and RITTER (1935). They obtained a value for k_{sD} of the order of 10^{-17}. Later, a number of investigators made more experiments using very similar methods. Some of them also arrived at the order of 10^{-17}; others obtained considerably smaller values. It turns out that even the dissolution of dolomite is a very sluggish process and that it is difficult to decide whether and when equilibrium is attained. The difficulties are similar to those experienced with magnesite and were discussed under B.III.2. The lowest value for the solubility product of dolomite ($k_{sD} = 10^{-19.33}$) was derived by GARRELS, THOMPSON and SIEVER (1960) from solution experiments and a definition of equilibrium based on extrapolation. The determinations by other authors have been re-viewed by Hsu (1967). They cluster around a value slightly higher than 10^{-17}.

In this situation, a confirmation from thermochemical data which are independent from the solution behavior would be desirable. However, the data available are not complete enough to yield a very precise value. In particular, reliable determinations of the heats of solution for dolo-mite and related carbonates seem to be lacking (STOUT and ROBIE, 1963). The value for the solubility product of $10^{-18.6}$ which may be calculated from existing thermochemical data, as compiled by ROBIE and WALD-BAUM (1968), is therefore, characterized by a large margin of error, on the order of one unit of the exponent. Nevertheless, the value is in the range of the direct determinations based on dissolution experiments.

A rough estimate may be deduced from crystal ("lattice") energy, although the exact value has not yet been calculated for the dolomite structure. In view of the existing structural similarities, it is reasonable to assume a crystal energy intermediate between 701 and 771 kcal/mole, the values of LENNARD-JONES and DENT (1927) for calcite and magnesite, respectively. The average solubility product of dolomite $k_{av} = \sqrt{k_{sD}}$ (18a) should then also be intermediate between the products of magnesite ($10^{-8.1}$) and calcite ($10^{-8.35}$). The exact value of k_{av} is perhaps closer to $10^{-8.35}$ because one half of the cations in dolomite are Ca^{2+} ions. On the average, the cations in dolomite are thus less hydrophilic than Mg^{2+} ions, and the tendency of the cations to become dissolved should be less pronounced in dolomite than for the uniform Mg^{2+} population in mag-nesite. Furthermore, the ordering of the cations in dolomite should in-crease the stability of the crystal structure. An average solubility product slightly greater than $10^{-8.35}$ is equivalent to a true solubility product k_{sD} greater than $10^{-16.7}$. The above reasoning gives more weight to the k_{sD} values being on the order of 10^{-17} than to the smaller value near 10^{-19} of GARRELS, THOMPSON and SIEVER (1960).

From his solubility measurements in concentrated salt solutions HORN (1969) extrapolated $k_{sD} = 10^{-18.22}$ for zero ionic strength. HORN's value was chosen for the preparation of Fig. 50 not only because it lies about in the middle of the range of published k_{sD} values, but also because the solubilities of brucite and magnesite were determined by HORN ac-cording to the same method. These values are contained also in Fig. 50. The results of this one author were used in order to obtain an internally consistent diagram.

The exact value of the solubility product is of no great consequence as far as the stability of the mineral in sea water is concerned. GARRELS and THOMPSON (1962) determined the activities of the most important ionic species in sea water, which are listed in Table 26. From these data

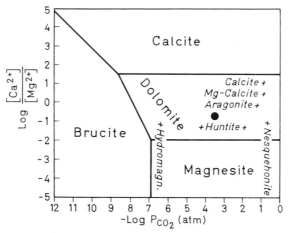

Fig. 50. Phase diagram of the system CaCO$_3$–MgCO$_3$–H$_2$O at 25° C based on the solubility products determined by HORN (1969) for zero ionic strength. $k_{s\,Brucite}$ $= 10^{-11.07}$, $k_{s\,Magnesite} = 10^{-8.14}$, $k_{s\,Dolomite} = 10^{-18.22}$, ($k_{s\,Calcite} = 10^{-8.35}$ after GARRELS et al., 1960). Phase boundaries are shown for these stable phases only. The metastable phases occurring in the system are indicated in italics in those regions of the diagram which approximately correspond to their conditions of formation. ● shows the composition of sea water

the activity product $[Ca^{2+}] \cdot [Mg^{2+}] \cdot [CO_3^{2-}]^2 = 10^{-15.01}$ is calculated. Sea water is thus considerably supersaturated with regard to dolomite, even for a solubility product in the order of 10^{-17}. Also the projection point of the ratio $[Ca^{2+}]/[Mg^{2+}]$ for sea water falls into the stability field of dolomite in Fig. 50. A value of 10^{-17} would result in a narrower dolomite field, but this would still contain the projection point for sea water. Dolomite should not only precipitate spontaneously from sea water, but also any immersed calcium carbonate should become converted to dolomite according to reaction (15), a conclusion which had been put forward by HALLA and RITTER (1935).

Judging from the activity product $[Mg^{2+}] \cdot [CO^{2-}] = 10^{-7.10}$, as calculated from the data in Table 26, sea water is even supersaturated about tenfold with regard to magnesite (cf. CHRIST and HOSTETLER, 1970). This property of sea water may be taken as the most striking illustration of the magnesite problem as discussed under B.III.2. In that section, the property of aqueous solutions to sustain considerable supersaturations with regard to anhydrous magnesium salts, notably magnesite, was explained by the strength of the H$_2$O—Mg^{2+} bond and the reluctance of the magnesium to release part or all of the water molecules which are attached to the ion in aqueous solutions and in hydrated crystals. This property of the magnesium ion must be responsible as well for the failure of dolomite to form from sea water despite high supersaturations. It is sufficient to recall the interpretation given on p. 114f. for the formation of aragonite under the influence of dissolved magnesium. The inhibitions opposing the formation of dolomite from sea water and similar supersaturated solutions are already explained in that section.

Another factor contributing to the inhibition may be seen in the disproportionate concentrations (molalities) in sea water of the component ions of dolomite. In particular, CO_3^{2-} is extremly low in comparison to the cations (see Table 26). The conclusion deduced from the kinetic model sketched in Fig. 32 has been that the growth of magnesite is favored by higher CO_3^{2-} concentrations. This should be valid also for dolomite where the problem of dehydrating the magnesium ions adsorbed on the surface of a nucleus is the same. The extremely low concentration of CO_3^{2-} in sea water will thus enhance the inhibitions due to the dehydration barrier, even though the dearth of CO_3^{2-} is overcompensated by high Mg^{2+} and Ca^{2+} in so far as the solubility product is concerned.

It must be borne in mind that it is not the total concentration of CO_3^{2-} which determines the supersaturation (and also the growth rate), but the effective concentration, i.e. the activity. This is still much lower $(4.7 \cdot 10^{-6})$ than the molality $(2.69 \cdot 10^{-4})$. This drastic lowering is due only in part to the low activity coefficient (of 0.20) of the CO_3^{2-} ion in sea water. The low activity is mainly caused by the tendency of CO_3^{2-} to form complexes with various cations, such as Ca^{2+}, Mg^{2+}, and Na^+. A considerable part (67%) of the total CO_3^{2-} is tied up with Mg^{2+} to form the neutral species $MgCO_3^0$. The equilibrium constant

$$\frac{[Mg^{2+}] \cdot [CO_3^{2-}]}{[MgCO_3^0]} = 10^{-3.4} \tag{19}$$

was determined at 25° C by GARRELS, THOMPSON and SIEVER (1961). It is quite close to the original value of GREENWALD (1941) who first drew attention to the importance of this type of complex formation. Taking account of equilibrium (19) and of similar complexing reactions, notably among the cations Ca^{2+}, Mg^{2+}, Na^+, and the anions CO_3^{2-}, HCO_3^-, SO_4^{2-}, GARRELS and THOMPSON (1962) calculated a chemical model for sea water at 25° C which is characterized by the values quoted in Table 26.

The occurrence of $MgCO_3^0$ is important not only in normal sea water, but also in sea water concentrated by evaporation. Due to the removal of Ca^{2+} from solution by the precipitation of aragonite and gypsum, concentrated brines possess a higher Mg/Ca ratio than normal sea water. For this reason, evaporation is regarded by many authors as a factor

Table 26. A chemical model for sea water at 25° C, chlorinity 19‰ and pH 8.1 from GARRELS and THOMPSON (1962) (see also GARRELS and CHRIST)

Ion	Molality (total)	% free ion	Activity coeffic.	Activity	Ion
Ca^{2+}	0.0104	91	0.28	0.00264	Ca^{2+}
Mg^{2+}	0.0540	87	0.36	0.0169	Mg^{2+}
Na$^+$	0.4752	99	0.76	0.356	Na$^+$
K$^+$	0.0100	99	0.64	0.0063	K$^+$
Cl$^-$	0.56	100	0.64	0.3584	Cl$^-$
SO$_4^{2-}$	0.0284	54	0.12	$1.8 \cdot 10^{-3}$	SO$_4^{2-}$
HCO$_3^-$	0.00238	69	0.68	$9.75 \cdot 10^{-4}$	HCO$_3^-$
CO$_3^{2-}$	0.000269	9	0.20	$4.7 \cdot 10^{-6}$	CO$_3^{2-}$

Resulting ionic activity products	Solubility products for comparison
$[Ca^{2+}] \cdot [CO_3^{2-}] = 10^{-7.91}$	$k_{sArag} = 10^{-8.22}$
$[Mg^{2+}] \cdot [CO_3^{2-}] = 10^{-7.10}$	$k_{sMagn} = 10^{-8.14}$
$[Ca^{2+}] \cdot [Mg^{2+}] \cdot [CO_3^{2-}]^2 = 10^{-15.01}$	$k_{sDolo} = 10^{-17}$

Data of comparable authority are not yet available for sea water under different conditions, e.g. for increasing depth. As evidenced by articles in the current literature, various authors obtain different results for the decrease of the ionic activity products with increasing depth. The situation is also complicated by different estimates of the increase of the solubility products with pressure.

essential in promoting dolomite formation. However, it must be doubted that the supersaturation with respect to dolomite is greatly increased by by evaporation. It is true that Mg^{2+} is increased, but also Ca^{2+} up to the point where gypsum precipitates, whereas the activity of CO$_3^{2-}$ must be expected to decrease by evaporation. In spite of a certain supersaturation with regard to aragonite, the ionic product $[Ca^{2+}] \cdot [CO_3^{2-}]$ will remain essentially constant due to the precipitation of aragonite. This process tends to remove CO$_3^{2-}$ from the evaporating sea water, so that the total CO$_3^{2-}$ will remain of the same order or may even decrease. At the same time, Mg^{2+} (also Na$^+$) is increased by the evaporation. Consequently, more CO$_3^{2-}$ will be captured by Mg^{2+} (and Na$^+$) to form the complex MgCO$_3^0$ (and NaCO$_3^-$). The amount of the free CO$_3^{2-}$ ion is 9% of the total CO$_3^{2-}$ in normal sea water, and this already small proportion will decrease further on evaporation. Although no exact calculations comparable to those of GARRELS and THOMPSON (1962) for normal sea water are available for brines concentrated and otherwise modified by evaporation, it may be estimated that the activity product $[Ca^{2+}] \cdot [Mg^{2+}] \cdot [CO_3^{2-}]^2$, i.e. the supersaturation with respect to dolomite, is not essentially increased by evaporation, but may even decrease in the process.

It is true that the molar ratio Mg/Ca may be increased by evaporation, but the ratio of 5.2 in normal sea water (activity ratio = 6.4) should be quite sufficient for reaction (16) to proceed in the direction of dolomite formation. If magnesium-bearing solutions would spontaneously equilibrate with immersed calcium carbonate according to (15), then an activity ratio $[Mg^{2+}]/[Ca^{2+}]$ exceeding $0.03 = 10^{-1.5}$ would be sufficient for dolomitization to proceed. This can be seen in Fig. 50, for which the solubility product of dolomite $k_{sD} = 10^{-18.22}$ has been assumed. Even if $k_{sD} = 10^{-17}$ were the correct value, an activity ratio of 0.5 would be sufficient, i.e. a figure about ten times smaller than the one characteristic of normal sea water.

We will no longer attach too much importance to elevated Mg/Ca ratios caused by evaporation when we know about the decrease of free CO_3^{2-} which accompanies the concentration of magnesium as a consequence of equilibrium (19). A high Mg/Ca ratio is another way of describing a high concentration of magnesium in sea water, and both descriptions are equivalent as long as evaporation and ensuing precipitations are the only processes modifying the composition of sea water. In this case, an elevated Mg/Ca ratio will be no more conducive to dolomite formation than high concentration of Mg^{2+}. On the contrary, let us consider the kinetic model for the growth of magnesite suggested by Fig. 32 and adopt it with due modifications for the case of dolomite. Then we see that the decrease of the free CO_3^{2-} anion will further enhance the inhibitions which already prevent dolomite formation in normal sea water.

However, it will be shown later that elevated Mg/Ca ratios are indeed likely to promote dolomitization provided that the increase is caused not simply by evaporation, but by the presence of high $[CO_3^{2-}]$ in solution, which depresses the calcium concentration by virtue of the ionic product $[Ca^{2+}] \cdot [CO_3^{2-}]$.

4. Magnesian Calcites in Aqueous Systems

From extrapolation of the diagram in Fig. 49 it was concluded that magnesian calcites can exist only in a metastable state at earth-surface temperatures. This view is confirmed by the solution behavior of magnesian calcites. As mentioned in connection with the solubility product of dolomite, the simplest method of determining the solubility of a sparingly soluble carbonate involves the use of (distilled) water, which is saturated at one atmosphere by bubbling with CO_2. The system is governed by the equilibria (9) and (10) (see p. 103), which may be combined by eliminating $[H_2CO_3]$. For one atmosphere CO_2 we have at $25°\,C$

$$[HCO_3^-] \cdot [H^+] = 10^{-7.82} \tag{20}$$

or using logarithms

$$\log[HCO_3^-] = pH - 7.82 \,. \tag{20a}$$

Before any carbonate is dissolved, the HCO_3^- ions are completely balanced by protons, i.e. the concentrations of HCO_3^- and H^+ are equal and their activities approximately so. Thus the pH of distilled water saturated with CO_2 at one atmosphere is about 4. By the dissolution of a carbonate, more HCO_3^- will be present in solution to balance the positive charges of the dissolved cations, i.e. to secure electroneutrality. Consequently, the pH rises according to (20a). At the same time, CO_3^{2-} increases by the resulting shift of equilibrium (11) (see p. 103) and limits the amount of dissolved carbonate by virtue of the solubility product $[X^{2+}] \cdot [CO_3^{2-}] = k_{sX}$. In this manner, the rise of the pH is limited as well. The solubility product can now be calculated from the equilibrium pH by taking account of (11), (20), and the electroneutrality condition. The calculation is straightforward when the activities are assumed equal to the concentrations as a first approximation. This may be used as a starting point for determining the activity coefficients by trial and error. Several cycles of refinement are usually needed for the exact calculation of the solubility product in terms of activities. A sample calculation for the case of dolomite has been published by GARRELS et al. (1960) (see also GARRELS and CHRIST, pp. 74—89).

For the purpose of merely comparing the solubilities of different carbonates, the solubility products need not be calculated in every case. According to the interrelations just pointed out, the final pH increases with the concentration of the cations dissolved from the carbonate studied. The final pH is approximately a linear function of the logarithm of the solubility products for simple carbonates and of suitably defined average products [see e.g. (18a) for the case of dolomite] for complex carbonates [GARRELS et al. (1960), Fig. 3]. For different carbonate materials tested in dissolution experiments as sketched above, the final pH values thus afford a scale of relative solubilities which should be essentially independent of chemical composition, in particular, when carbonates containing the same cations are compared.

CHAVE, DEFFEYES, WEYL, GARRELS and THOMPSON (1962) have determined the final pH values at one atmosphere CO_2 pressure for a number of skeletal carbonates, notably magnesian calcites of different composition. Below about 5 mol% Mg they found the same pH as for pure calcite (pH = 6.02). More highly magnesian calcites yield higher values for the final pH. The maximum pH of 6.35 was found for 25 mol% Mg. Magnesian calcite with 10 mol% Mg is about as soluble as aragonite, which is characterized by an equilibrium pH of 6.08.

JANSEN and KITANO (1963) also observed the general tendency of magnesian calcites to be more soluble with increasing magnesium content. They appear to have worked at the partial pressure of the surrounding air ($P_{CO_2} = 10^{-3.5}$ atmospheres). Their results are thus not directly comparable with those of CHAVE et al. (1962) who worked at one atmosphere CO_2. In addition to pH measurements, JANSEN and KITANO subjected their solutions to chemical analysis. In distilled water they found that only dissolved magnesium increased with the magnesium content of the calcites tested, whereas the final concentration of calcium was found to be independent of the Mg mole percentage. The authors give evidence suggesting that essentially pure calcium carbonate was being reprecipitated within one week, which must be the reason for the enrichment of Mg^{2+} in solution.

This mechanism was also discussed by BERNER (1967). Consequently, the final or "steady-state" pH does not constitute an absolute measure for the solubility of magnesian calcites. It can also vary with reaction time and the pretreatment of the specimens, e.g. the intensity of grinding. Even so, for uniform experimental conditions, the method does afford at least a relative scale of solubility. Furthermore, it is demonstrated that magnesian calcites are metastable in fresh water with respect to pure calcite and that the more highly magnesian calcites (> 10 mol%) are even less stable than aragonite. This result is in harmony with mineralogical observations made in marine carbonates undergoing fresh-water diagenesis, where magnesian calcites tend to become converted to calcite more rapidly than aragonite (STEHLI and HOWER, 1961; FRIEDMAN, 1964; GAVISH and FRIEDMAN, 1969).

JANSEN and KITANO carried out additional dissolution experiments involving various salt solutions. The most interesting result is that $MgCl_2$ with the same concentration as in sea water (0.052 M) does not depress the solubility of magnesian calcites, but rather increases it considerably. This result may seem surprising at first sight when considered exclusively from the point of view of the solubility product. However, we have to take account of the complex formation of Mg^{2+} with CO_3^{2-} according to (19), which lowers the amount of free CO_3^{2-} and thus enhances the tendency of any carbonate to dissolve. Obviously this mechanism overcompensates the direct influence of more Mg^{2+} on the solubility. Contrary to what one would expect and different from aragonite, magnesian calcites are thus not "stabilized" by dissolved Mg^{2+} alone.

11*

NaCl, when present in solution at about the same level (0.405 M) as in sea water also increases the solubility of magnesian calcites. Again this is to be explained by CO_3^{2-} being tied up in complexes of the composition $NaCO_3^-$ (GARRELS et al., 1961).

The solubility is depressed considerably by $NaHCO_3$ added at the HCO_3^- concentration level of sea water (0.0023 M). The HCO_3^- acts through the CO_3^{2-} ions with which it is accompanied in solutions, according to equilibrium (11). It can be concluded that the small amount of free CO_3^{2-}, buffered by the HCO_3^- present in sea water, is responsable for the persistence of magnesian calcites in the marine environment. The CO_3^{2-} acts of course in combination with the Mg^{2+} and Ca^{2+} of the sea water.

The most important general result of JANSEN and KITANO is the demonstration that the final solutions in almost all cases contained Ca^{2+} and Mg^{2+} in proportions which are at variance with the compositions of the magnesian calcite dissolved. This result was explained by the reprecipitation of calcium carbonate(s). Thus the dissolution of magnesian calcites is characterized by a high degree of irreversibility, and attempts to derive a quantitative description of the instability, e.g. in terms of free energy, appear hopeless from solubility data.

LERMAN (1965) has calculated such data from the high-temperature phase diagram of GOLDSMITH and HEARD (1961) and he found the maximum free energy of mixing of 355 cal/mole for a magnesian calcite near 30 mol% $MgCO_3$. [The free energy difference between calcite and aragonite, which is of the order of 200 cal/mole (MACDONALD, 1956), may be mentioned for comparison.]

The irreversibility of the solution behavior indicates also that the conditions prevailing in the dissolution experiments must be far removed from the conditions of formation for magnesian calcites. This is true not only for the experiments of CHAVE et al. (1962) who worked at one atmosphere CO_2 pressure with resulting pH values of the order of 6 to 7, but also for those of JANSEN and KITANO, who worked at pH values between 8 and 9. In the same pH range, a maximum of about 5 mol% $MgCO_3$ was incorporated into some of the calcites obtained in the experiments summarized in Tables 23 and 24.

So far, only in the experiments of KITANO and KANAMORI (1966), where organic additives were used, were more highly magnesian calcites obtained at pH values between 8 and 9 and at relatively low concentrations of Mg^{2+} and Ca^{2+} on the order of 0.00X M. It was already mentioned on p. 132 that the citrate, the pyruvate, and the malate of sodium were the essential cosolutes in these syntheses, in which the method of CO_2 escape was used. The maximum contents of $MgCO_3$ approached 15 mol% for additions of 2 to 3 grams per liter of the organic radicals.

In the absence of organic cosolutes, i.e. in systems involving Ca^{2+}, Mg^{2+}, CO_3^{2-}, and HCO_3^- plus inorganic balancing ions, magnesian calcites of similar composition were obtained by other authors as precipitates from more concentrated solutions. ERENBURG (1961) mixed 1/6 M solutions of $CaCl_2$ and $MgCl_2$ with a solution which was 0.5 M for both Na_2CO_3 and $NaHCO_3$ (estimated pH = 10). By precipitating at 50° C and varying the relative amounts of $CaCl_2$ and $MgCl_2$, he obtained a series of magnesian calcites ranging from 17 to 55 mol% $MgCO_3$, for which he determined the lattice constants. In the compositional range near 50 mol% no superstructure reflections were present, so that these products may be referred to as disordered dolomites. SIEGEL (1961) also obtained magnesian calcites near 50 mol%, at 25°, 50°, and 100° C, by coprecipitating 1.0 M and 0.5 M solutions of $Ca(NO_3)_2$, $MgSO_4$, and Na_2CO_3 (estimated pH near 12), which he used in the concentration ratios of 1:1:1 and 1:1:2.

GLOVER and SIPPEL (1967) carried out precipitations in the range of 0° to 40° C at rather high concentrations. They used $NaHCO_3$ at the initial concentration of 0.3 M (estimated initial pH = 8.4) as the source for the carbonate. Their experiments were thus not pure precipitations but included the method of CO_2 escape. By varying the initial $MgCl_2$ concentration from 0.08 M to 1.3 M and with fixed 0.05 M $CaCl_2$, they obtained magnesian calcites containing up to over 30 mol% $MgCO_3$. Magnesium contents of over 50 mol% in the final product were obtained in the presence of high NaCl concentrations and by increasing the initial $MgCl_2$ level above the range quoted above.

If we set aside the experiments of KITANO and KANAMORI, which involve additions of organic reagents and which may perhaps serve as models for the formation of magnesian calcites in organisms, then we may conclude from the experiments of ERENBURG and SIEGEL that high pH values promote the incorporation of magnesium into rhombohedral carbonate structures. However, the syntheses of GLOVER and SIPPEL, starting from $NaHCO_3$ with a pH of about 8.4, suggest that an extremely high pH is perhaps not as important as an ample supply of CO_3^{2-} ions. This interpretation would conform with the ideas developed above for the growth kinetics of magnesite. According to Fig. 32, sufficient supply of CO_3^{2-} ions should be the most likely way of overcoming the dehydration barrier, the most serious obstacle opposing the incorporation of magnesium ions into anhydrous carbonates.

In the experiments just reviewed, the supply of cations is obviously too excessive for cation ordering to take place or, in particular, for dolomite or huntite to form. The concentrations are also too high to justify direct comparisons with natural conditions of formation. Nevertheless, the condition that all component ions, including CO_3^{2-}, must be present in solution at comparable concentrations for magnesium-bearing rhombohedral carbonates to form, may be of more general importance, and is perhaps required also in the case of dolomitization. This view will find additional support on pp. 169—177. Finally the experiments of ERENBURG (1961) and GLOVER and SIPPEL (1967) have thus far offered the only possibility for covering the whole compositional range of magnesian calcites up to over 50 mol% $MgCO_3$ by syntheses from aqueous solutions.

5. "Dedolomitization"

The term dedolomitization has been used by many authors to denote the hypothetical reaction:

$$CaMg(CO_3)_{2\,(solid)} + Ca^{2+} \rightarrow 2CaCO_3 + Mg^{2+} \qquad (15a)$$

which is the reverse of reaction (15). Dolomite is supposed to convert to calcium carbonate, usually calcite, under the influence of solutions rich in calcium.

SMIT and SWETT (1969) have justly criticized the term on formal grounds for its ambiguity. Their main arguments are as follows: A process should be named by the replacing mineral, not the one replaced. The term is geochemically misleading because magnesium ions are supposed to be removed whereas the other components remain essentially in place. The authors, therefore, suggest instead to speak of calcitization of dolomite.

However, it is doubtful whether the process described by Eq.(15a) is of any consequence in nature. As pointed out on p. 156, the dissolution of dolomite is a sluggish process even at one atmosphere CO_2 pressure, i.e. under conditions where calcite is readily dissolved. It is therefore hard to imagine kinetic conditions for which the opposite is true, even though the thermodynamic requirements for reaction (15a) may be met. No solution experiments seem to have been carried out so far on disordered magnesium-deficient dolomites. In the same way that magnesian calcites are more soluble than calcite, such calcian dolomites should be more soluble than dolomite according to the extrapolation of the phase diagram in Fig. 49. In view of the difficulties experienced with the solubility determination for dolomite, and also for more highly magnesian calcites, it may be hard to prove the higher solubility of calcian dolomites by direct experiments.

However, KATZ (1968) found that what is customarily referred to as dedolomitization affects only calcian dolomites, at least in the samples he studied. It is conceivable that calcian dolomites above a certain excess of calcium are more soluble even than calcite. In this manner, the process interpreted as dedolomitization may consist actually of the preferential dissolution of calcian dolomites accompanied by the reprecipitation of calcium carbonate. This situation would be similar to the behavior of magnesian calcites in the solution experiments of JANSEN and KITANO (see p. 163).

The same interpretation will probably apply to other examples of alleged dedolomitization. However, in most pertinent publications the composition and the state of order are not determined for the dolomites involved. Rhombohedral voids in the centers of dolomite rhombohedra (BAUSCH, 1965; MATTAVELLI, 1966) may have originated from massive dolomite crystals with more highly calcian cores, which were dissolved preferentially at some stage of diagenesis.

6. Huntite

In their experiments at one atmosphere CO_2, GARRELS et al. (1960) determined a final pH of 6.24 for huntite. This corresponds to an "average" solubility product of about $10^{-7.8}$. Huntite is thus metastable with respect to mixtures of magnesite and dolomite or calcite. It may be suggested that huntite forms in nature in spite of its metastability for kinetic reasons. It is possible that its comparatively open structure (p. 43) is especially prone to facilitate the dehydration of Mg ions during crystal growth.

IV. The Aqueous Synthesis of Norsethite, $BaMg(CO_3)_2$, a Model for Low-Temperature Dolomite Formation

1. General Statement

The preceding chapters on the system $CaCO_3–MgCO_3$ may be summarized as follows. Dolomite must have a large stability field at earth-surface temperatures (see Fig. 50), notably in the marine environment. This is in harmony with a great many field observations attesting to the low-temperature origin of the mineral in marine formations and its persistence during geologic periods of time. By virtue of its solubility product, dolomite should easily form from sea water and related solutions, both natural and artificial. The fact that dolomite is not a normal product of marine carbonate precipitation (and of many experiments) implies the existence of serious kinetic obstacles. The most important inhibition must be due to the dehydration barrier of the Mg^{2+} ion. This concept was introduced on pp. 76—85 in order to explain the failure of magnesite to crystallize at room-temperature. Since magnesium ions must be dehydrated before they can be incorporated into any anhydrous magnesium salt, the dehydration barrier will inhibit the formation from aqueous solutions of dolomite as well (see p. 114f. and p. 158f.).

However, from this theory nothing more than suggestions may be derived as to the conditions which actually lead to dolomite formation in nature. Elevated $[CO_3^{2-}]$ in solution was deduced from the model sketched in Fig. 32 as a possible factor promoting the crystallization of anhydrous magnesium-bearing carbonates. More straightforward information cannot be expected from the system $CaCO_3–MgCO_3$ as long as it is not possible to synthesize dolomite under its apparent conditions of formation in nature.

From the viewpoint of the dehydration barrier, the successful synthesis of any other anhydrous carbonate containing magnesium as an essential constituent should contribute valuable clues towards a solution of the dolomite question. More general evidence that the special behavior of the magnesium ion is responsible for the failure to prepare dolomite below 100° C comes from a comparison of the crystal structures of dolomite and related simple carbonates. On the one hand, dolomite may be described as a regular interstratification of calcite-like and magnesite-like layers (p. 23f.). On the other hand, in contrast to calcite, magnesite and dolomite have not so far been prepared from aqueous solutions at room-temperature. The common denominator appears to be the magnesite layer, the building of which must be the rate-determining step inhibiting the crystallization of magnesite and dolomite.

On p. 36, norsethite, $BaMg(CO_3)_2$, was shown to contain layers of magnesium carbonate which are very similar to the ultimate layers of magnesite and to the magnesite-like layers in dolomite (see Fig. 51). Besides the possibility of preparing the mineral at high temperatures and

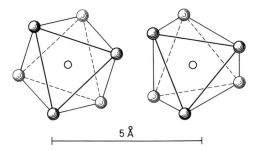

5 Å

Fig. 51. Comparison of MgO_6 octahedra in dolomite and norsethite. Projections on the basal plane along the hexagonal c axis. The octahedron in dolomite (left side) deviates from cubic symmetry by a dilatation in the direction of the c axis, which does not show in this projection. – The octahedron in norsethite (right side) is distorted in addition by a mutual twist of its basal triangles

pressures (500° C and 15 kb; CHANG, 1964), norsethite can also be synthesized in aqueous solutions at room-temperature (LIPPMANN, 1967, 1968 b). This synthesis may help to elucidate the conditions of dolomite formation not only by virtue of existing chemical and structural analogies. In addition, the crystallization of norsethite is the only known low-temperature reaction which allows the stoichiometric incorporation of large amounts of magnesium into an anhydrous carbonate within reasonable periods of time.

2. Experimental Norsethite Formation

The description of the synthesis experiments may be introduced by the statement that $BaMg(CO_3)_2$ does not form directly as a precipitate from solutions which contain all three component ions in sizable amounts. Barium carbonate would be the initial product under these conditions. The situation is reminiscent of the conditions of formation as generally postulated for dolomite, which has been assumed to form chiefly by reaction of preexisting calcium carbonate sediments (cf. p. 149). In an analogous fashion, sizable amounts of norsethite are obtained only by reaction of solid barium carbonate with appropriate magnesium-bearing solutions. However, the process does not follow the analogue of reaction (15):

$$2\,BaCO_{3\,(solid)} + Mg^{2+} \rightarrow BaMg(CO_3)_{2\,(solid)} + Ba^{2+}. \qquad (21)$$

Barium carbonate may be kept immersed in magnesium chloride solutions of various concentrations for months and years without any observable sign of the anticipated replacement reaction. What can be observed in the more concentrated $MgCl_2$ solutions (above about 0.1 M) is the dissolution of $BaCO_3$ when added in the amount of about 200 milligrams per liter. This solvent action may be attributed to the removal of free CO_3^{2-} ions from solution by the formation of $MgCO_3^0$ complexes according to equilibrium (19):

$$BaCO_{3\,(solid)} + Mg^{2+} \rightarrow MgCO_3^0 + Ba^{2+}. \qquad (22)$$

Norsethite forms within a few days, when solutions are used which contain carbonate ions in addition to magnesium chloride. Additions of bicarbonate, e.g. $NaHCO_3$, serve the same purpose. It must be remembered that bicarbonate is always accompanied in solution by carbonate ions due to the reaction:

$$2\,HCO_3^- \rightarrow CO_3^{2-} + H_2CO_3 \qquad (23)$$

which is governed by the equilibria (cf. p. 103)

$$\frac{[H^+] \cdot [HCO_3^-]}{[H_2CO_3]} = 10^{-6.35}, \tag{10}$$

$$\frac{[H^+] \cdot [CO_3^{2-}]}{[HCO_3^-]} = 10^{-10.33}. \tag{11}$$

The pH value buffered by the solution of a bicarbonate may be estimated by assuming activities equal to concentrations. When no exchange of CO_2 takes place with the atmosphere, equal amounts of CO_3^{2-} and H_2CO_3, as produced by (23), are present in the solution. By combining (10) and (11), a value of 8.34 is then obtained as a first approximation for the pH of the solution of a bicarbonate. Substituting this result in (11) we see that dissolved bicarbonate is accompanied by carbonate of about 10^{-2} times the concentration of the bicarbonate.

The solution from which the present writer obtained his first norsethite contained 0.01 M $MgCl_2$ along with 0.02 M $NaHCO_3$. Barium carbonate *pro analysi* was added to the solution. This reagent yields essentially the same X-ray powder pattern as the mineral witherite and consists of very small acicular crystals, elongated in the c direction, up to about 10 microns long and 2 microns thick. By reaction of this material with the solution the first crystals of norsethite become noticeable after two days. They are more equant than the particles of the starting material and attain sizes on the order of 100 microns when allowed to continue their growth. They usually cling to the bottom and the walls of the vessel used. This behavior is helpful in recognizing the first appearance of the crystals and in separating them from unreacted barium carbonate, most of which remains loose.

The identity of the reaction product with norsethite follows from its optical properties as determined under the petrographic microscope (LIPPMANN, 1968a), the X-ray powder pattern (see Fig. 13) and (unpublished) quantitative chemical analyses. A molar ratio Mg/Ba near 1:1 could also be inferred from the presence of the superstructure reflections of the type $(0k.l)$, with l odd (see p. 32 and Fig. 13). The crystal structure determination (p. 32ff.), which was carried out for such artificial norsethite, affords an additional confirmation of the stoichiometric composition $BaMg(CO_3)_2$ and attests to the well-ordered character of the product.

In a closed vessel, the "norsethitization" of witherite by the above specified solution remains incomplete even over periods of months, and in spite of a stoichiometric excess of $MgCl_2$ and $NaHCO_3$ with respect to the amount of $BaCO_3$ used. It turns out that in the course of six months the pH has dropped from the starting value of 8.3 to 7.5. Considering equilibrium (11), we come to the conclusion that the solution must have been depleted of CO_3^{2-} ions. The norsethitization may be reactivated by leaving the vessel open for some time. These observations show that CO_3^{2-} is actually consumed in the formation of norsethite. Evidently CO_2 escapes according to

$$H_2CO_3 \rightarrow H_2O + CO_2 \quad \text{[governed by equilibrium (9) p. 103]}$$

so that more CO_3^{2-} can be produced by reaction (23).

The simplest overall reaction involving consumption of CO_3^{2-} ions is:

$$BaCO_{3\,(solid)} + Mg^{2+} + CO_3^{2-} \rightarrow BaMg(CO_3)_{2\,(solid)} . \qquad (24)$$

Other reactions leading from witherite to norsethite could be imagined. Since norsethite is well-defined stoichiometrically by the formula $BaMg(CO_3)_2$, such reactions could only consist of linear combinations of (21) and (24). However, a (21) component would set Ba^{2+} free, and this would be removed immediately from solution, either as witherite or as norsethite by the CO_3^{2-} ions which must be present for norsethitization to proceed. Thus (24) is the only consistent overall reaction by which both Mg^{2+} and CO_3^{2-} are consumed in the process of norsethitization.

$PbMg(CO_3)_2$, a new rhombohedral double carbonate (LIPPMANN, 1966) may be prepared in the laboratory at room-temperature in much the same way as norsethite, even though at considerably slower rates. Reagent grade $PbCO_3$, which yields the same X-ray powder pattern as the mineral cerrussite, is immersed in a solution containing 0.01 M $MgCl_2$ along with 0.02 M $NaHCO_3$. The first crystals of $PbMg(CO_3)_2$ may be observed after some months. In the course of half a year, they have grown to a size of about 20 microns. The crystal forms, as seen under the microscope, may be interpreted as trigonal trapezohedra. This symmetry and a calculated structure model based on maximum electrostatic stability (LIPPMANN, 1968a) are the main criteria for assigning $PbMg(CO_3)_2$ to the structure type of norsethite. Due to the slow rate of crystallization, it is difficult to observe a definite standstill of the $PbMg(CO_3)_2$ formation in a closed vessel as in the case of norsethitization. However, in the course of 15 months, the pH was found to have decreased from an initial value of 8.4 to 7.9. On the same grounds as in the case of norsethitization, this trend shows that CO_3^{2-} ions are consumed in addition to Mg^{2+} in the process of $PbMg(CO_3)_2$ formation from $PbCO_3$. The type of reaction (24) is thus not restricted to norsethitization, and we may write an analogous overall reaction for the synthesis of $PbMg(CO_3)_2$:

$$PbCO_{3\,(solid)} + Mg^{2+} + CO_3^{2-} \rightarrow PbMg(CO_3)_{2\,(solid)} . \qquad (25)$$

However, in view of the faster rate of (24) in comparison to (25), more evidence confirming the type of reaction may be expected mainly from experiments involving barium. A direct proof of (24) is obtained by adding the magnesium not as the chloride, as described above, but as a carbonate. In a 0.02 M solution of $NaHCO_3$ a given amount of $BaCO_3$ may be norsethitized completely within a few weeks by $MgCO_3 \cdot 3\,H_2O$ (artificial nesquehonite) added in slight excess over the stoichiometrically required amount according to (24). Different from experiments involving $MgCl_2$, the reaction does not come to a standstill in a closed vessel. The dissolution of nesquehonite supplies both the Mg^{2+} and the CO_3^{2-} so that the solution is not preferentially depleted of CO_3^{2-} as in closed-vessel experiments starting from $MgCl_2 + NaHCO_3$.

Norsethitization of $BaCO_3$ may be effected in distilled water in the absence of other dissolved compounds when the required amount of magnesium is added as basic magnesium carbonate, which yields essentially the same X-ray powder pattern as the mineral hydromagnesite. The vessel was allowed to stand open occasionally in order to allow the absorption of CO_2 from the air. This supply is obviously needed to compensate for the CO_3^{2-} deficit resulting from the presence of OH^- as an essential anion in hydromagnesite.

Reaction (24) may be confirmed also by replenishing the consumed ions in small steps by adding solutions of appropriate concentrations from pipettes. The solutions to be added must contain more reagent than is actually consumed, in order to prevent the dilution of the solution in the reaction vessel. When two different solutions are added, they must contain double the concentration to be maintained in the reaction vessel plus the amount of reagent consumed by (24). For the preparation of greater quantities of norsethite, it is advisable also to add the barium carbonate in small steps as a suspension from a pipette in order to minimize its occlusion in the reaction product. Two different replenishing solutions are sufficient to keep reaction (24) going. Carbonate ions are added as dissolved Na_2CO_3 by the first solution. In addition, this solution contains two times the concentrations in the reaction vessel of $NaHCO_3$ and Na_2CO_3 in order to keep constant the concentrations of bicarbonate and carbonate. At the same time, this replenishing solution may be used to introduce stoichiometric amounts of suspended $BaCO_3$. Pipettes of equal volumes must be used to add this suspension and, at the same pace, the second solution which contains the stoichiometric equivalent of $MgCl_2$ plus double the $MgCl_2$ concentration in the reaction vessel.

In the two replenishing solutions, the CO_3^{2-} and Mg^{2+} ions to be consumed in norsethitization are balanced by Na^+ and Cl^-, respectively. Therefore, the reacting solution should contain NaCl from the beginning at a concentration equivalent to the added concentrations of $BaCO_3$, Na_2CO_3, and $MgCl_2$ to be consumed according to (24). In this manner, the NaCl level in the reaction vessel will remain constant despite the continuous additions of Na^+ and Cl^- not participating in reaction (24).

By using the technique just outlined, considerable amounts of $BaCO_3$ can be converted to $BaMg(CO_3)_2$ in solutions of constant composition, provided that norsethitization proceeds exactly according to (24). Any deviations from the stoichiometry of (24) would be indicated by a gradual change of the pH in the course of the conversion of larger amounts of $BaCO_3$. For example, if the CO_3^{2-} equivalent added by the first replenishing solution were not completely incorporated into norsethite, then the pH in the reaction vessel would rise consistently. In the experiments to be specified below, a rise in pH could indeed be registered on addition of a sufficient amount of replenishing solution. However, after the material added had reacted, the pH returned to its initial value. No consistent increase (or decrease) was observed after a number of consecutive additions when these were timed according to the reaction rate as indicated by the disappearance of loose $BaCO_3$.

In addition to confirming reaction (24), the norsethitization of sizable amounts of $BaCO_3$ at constant concentration may serve to estimate reaction rates. Nucleation is not the main problem. It occurs spontaneously and its rate could be determined quantitatively, if desired. The overall rate of norsethite formation depends strongly on the amount of nuclei available. For the same solution composition, the rate is slow at first and increases with the amount of $BaCO_3$ converted in the reaction vessel. A steady state appears to prevail as soon as the bottom of the vessel is covered with a continuous crust consisting of norsethite crystals.

In this situation, about 5 milligrams of $BaCO_3$ (referred to a crust area of about 200 cm^2) may be converted per day in a solution containing:

$$0.01 \text{ M } MgCl_2; \quad 0.005 \text{ M } NaHCO_3; \quad (10^{-4} \text{ M } Na_2CO_3;)$$
$$0.02 \text{ M } NaCl; \quad pH = 8.6 \tag{a}$$

whereas about 5 milligrams of $BaCO_3$ per hour may be norsethitized in the following solution

$$0.003 \text{ M } MgCl_2; \quad 0.02 \text{ M } NaHCO_3; \quad 0.003 \text{ M } Na_2CO_3;$$
$$0.02 \text{ M } NaCl; \quad pH = 9.1. \tag{b}$$

There is thus no positive correlation of the norsethitization rate with the magnesium concentration; but the reaction is speeded by a higher concentration of CO_3^{2-}. A similar relationship appears to exist for the nucleation rates. Whereas in solution (a) the first crystals may be noticed after about a fortnight, they appear after a few days in solution (b).

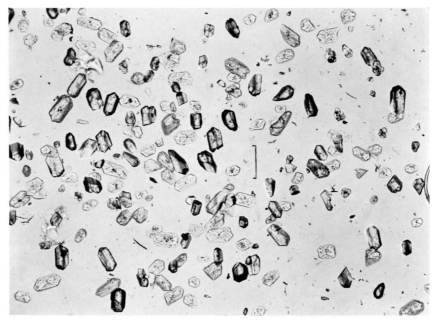

Fig. 52. Prismatic crystals of norsethite terminated by rhombohedral faces. They formed in solution (b). Plain polarized light vibrating in the direction of the scale; scale length: 0.1 mm; refractive index of immersion medium: 1.66. – The hour-glass structure shows in those crystals which exhibit n_ω, i.e. whose prisms are oriented (nearly) perpendicular to the scale

The norsethite crystals obtained in solution (b) are slightly inhomogeneous optically. The somewhat elongated hexagonal prisms terminated by trigonal rhombohedral faces (Fig. 52) contain an hourglass structure. The index $n_{\omega 1} = 1.693$ of the interior part, i.e. of the hourglass proper, coincides with the value normally observed in norsethite. A slightly lower index $n_{\omega 2} = 1.687$ is found for the remaining crystal. The crystals grown in solution (a) are steep rhombohedra (Fig. 53) which are optically homogeneous and show a uniform $n_\omega = 1.693$. Both types of crystals are uniform with respect to $n_\varepsilon = 1.517$ and cannot be distinguished by other criteria, such as X-ray diffraction and chemical composition (LIPPMANN, 1968a). Obviously the greater reaction rate in solution (b) is correlated with distinct deviations from the ideal crystal.

3. The Mechanism of Norsethitization

The experiments described above hint at a reaction via solution and attest to the eminent rôle played by dissolved CO_3^{2-} ions in the process of

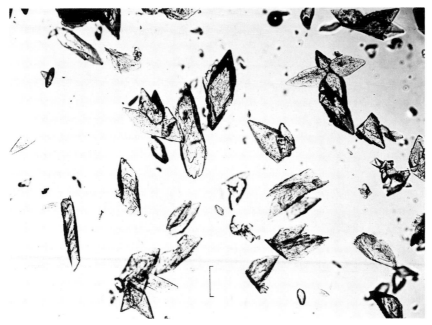

Fig. 53. Steep rhombohedra of norsethite grown in solution (a). Same scale and same optical conditions as in Fig. 52

norsethite formation. As in most room-temperature reactions involving solids, the starting material ($BaCO_3$) must first dissolve before its component ions can participate in further reactions leading to the crystallization of norsethite. The dissolution of barium carbonate is governed by the solubility product:

$$[Ba^{2+}] \cdot [CO_3^{2-}] = 10^{-8.6} . \tag{26}$$

That norsethitization is possible only in the presence of additional dissolved CO_3^{2-} and that the reaction is even accelerated by higher concentrations of CO_3^{2-} excludes the dissolution of $BaCO_3$ as a rate-determining step. [The rate of $PbMg(CO_3)_2$ formation might be determined by the dissolution of $PbCO_3$, since the solubility product of the latter, which from the literature is of the order of 10^{-13}, is extremely low.] From the viewpoint of barium, it is a paradox that it can react only when its concentration is depressed considerably, according to (26), by additions of dissolved CO_3^{2-}. The favorable effect of low Ba^{2+} can be understood, nevertheless, when the situation at the surface of a growing norsethite crystal is considered and when account is taken of ionic hydration.

The likelihood of Ba^{2+} ions being adsorbed at a cation site is much greater than for Mg^{2+} ions, which are more tightly hydrated (see Tables 20 and 21), resulting in very few being sufficiently activated to break their hydration envelope when impinging on a growing crystal. If barium and magnesium are present in solution at comparable concentrations, Ba^{2+} will occupy not only its normal sites in the crystal surface, but it may be expected to congest also the sites which should be filled by Mg^{2+} in a growing norsethite crystal. Obviously, the concentration of Ba^{2+} must be depressed to such an extent that the Mg^{2+} sites remain open long enough to be filled by one of the few activated Mg^{2+}. The ratio Mg/Ba may be estimated to be about 200 in solution (a) and 3000 in solution (b). The reciprocals of these figures may be regarded as indicating the order of magnitude of the fraction of activated Mg^{2+}, for the concentration of activated Mg^{2+} should be about equal to the total concentration of Ba^{2+} for crystal growth to proceed at an optimum rate. Of course, CO_3^{2-} also should be available in comparable amounts. Thus besides depressing Ba^{2+} according to (26), the CO_3^{2-} ions added to the solution as sodium (or magnesium) carbonate and/or bicarbonate are of more immediate importance in that they are also essential constituents of norsethite.

The situation at the surface of a norsethite crystal undergoing growth is similar to an assembly line, where all component parts have to be supplied in appropriate proportions. This condition is obviously not satisfied for a nucleus in contact with a $MgCl_2$ solution containing immersed $BaCO_3$, but devoid of additional carbonate. A shortage of one component part (CO_3^{2-}) and excessive rejects (insufficient activation) in another (Mg^{2+}) cannot be made up by the oversupply of a third component part (Ba^{2+}). This would only lead to congestion in the shop (crystal surface) and production (crystal growth) would break down.

However, this comparison is valid only to a certain extent. The fast reaction rate in solution (b) is caused by a distinct oversupply of CO_3^{2-}. This effect indicates that dissolved CO_3^{2-} plays a rôle in norsethitization in addition to that of depressing the Ba^{2+} concentration and of being an essential constituent of the final product. A promoting influence of high dissolved CO_3^{2-} in the crystallization of anhydrous magnesium carbonates follows from the kinetic model sketched in Fig. 32 for the case of magnesite growth from aqueous solution. Only sufficiently activated CO_3^{2-} will become bonded without intermediate H_2O, to Mg ions in the crystal surface, and the amount of such activated CO_3^{2-} increases with the total CO_3^{2-} concentration. This same mechanism has been suggested before (pp. 159, 165f., 168) as a possible way of overcoming the dehydration barrier of magnesium in connection with the formation of dolomite and magnesium calcites. The result that norsethitization is promoted by dissolved CO^{2-} may be regarded as an experimental confirmation of that hypothesis.

On p. 171 it was shown that norsethitization proceeds according to the overall reaction

$$BaCO_{3\,(solid)} + Mg^{2+} + CO_3^{2-} \rightarrow BaMg(CO_3)_{2\,(solid)}. \tag{24}$$

This equation adequately describes the material turnover as it is actually observed in the experiments. In addition, it contains information on the experimental conditions leading to the formation of norsethite: CO_3^{2-} must and Mg^{2+} may be present in solution at sizable concentrations, whereas the concentration of barium has to be limited by the sparing solubility of $BaCO_3$ and depressed further by dissolved CO_3^{2-} according to equilibrium (26).

However, from the explanation of the norsethitization mechanism as given in this chapter, it should be clear that Eq. (24) accounts for the changes in the system only on a megascopic scale. For $BaCO_3$ must first dissolve according to

$$BaCO_{3\,(solid)} \rightarrow Ba^{2+} + CO_3^{2-} \quad \text{[governed by (26)]} \tag{27}$$

in order for Ba^{2+} to be available for the ionic reaction. On the same grounds as in the case of dolomite formation and in analogy to Eq. (16), the true reaction leading to norsethite which describes the crystallization on the scale of the crystal structure, must be written:

$$Ba^{2+} + Mg^{2+} + 2CO_3^{2-} \rightarrow BaMg(CO_3)_{2\,(solid)}. \tag{28}$$

In opposition to Eq. (21) no Ba^{2+} appears on the right side of this equation, a feature which it has in common with the overall reaction (24). In point of fact, the only difference between (24) and (28) is the implication of the latter that $BaCO_3$ is already dissolved according to (27). Eq. (24) represents the simplest reaction which starts from solid $BaCO_3$ and which is, at the same time, closely related to the true ionic reaction (28).

We might also write Eq. (21) assuming that $BaCO_3$ has already dissolved:

$$2\,Ba^{2+} + 2\,CO_3^{2-} + Mg^{2+} \rightarrow BaMg(CO_3)_{2\,(solid)} + Ba^{2+}. \tag{21a}$$

We see that this form is different from the true ionic reaction (28) and is characterized by excess Ba^{2+}, which is not conducive to norsethite formation, as reported on p. 169. Obviously, norsethite does not form when all the CO_3^{2-} required is introduced in the form of the sparingly soluble $BaCO_3$.

4. Conclusions Regarding the Low-Temperature
Formation of Dolomite

From the failure to produce norsethite at room-temperature by the hypothetical reaction (21), it may be concluded that the analogous reaction (15) is perhaps equally inefficient in producing dolomite. As shown by USDOWSKI (1967), the process described by Eq. (15) does yield ordered dolomite above 120° C, when the Marignac reaction (14) is used. However, under unfavorable conditions, when larger crystals of calcite are reacted with MgCl$_2$ solutions, the product may consist of poorly ordered dolomite even if the experiment is carried out at a temperature as high as 200° C (BUBB and PERRY, 1968). The supply of Mg^{2+} to the unreacted calcite and the removal of the Ca^{2+} set free are possible only by diffusion through the pores of a crust of reacted material. Obviously, this exchange is not sufficient for reaction (15) to proceed in a stoichiometric fashion. The fact that some reaction does take place is indicative of an exchange sufficient to meet the thermodynamic requirements of reaction (15). The formation of poorly ordered dolomite, which is metastable, instead of ordered dolomite, which is the only stable intermediate phase in the system, then leads to the conclusion that the kinetic requirements of reaction (15) are hard to meet even at a temperature as high as 200° C. These difficulties must become more and more acute with decreasing temperature as evidenced by the failure to accomplish reaction (15) below 120° C.

A number of reasons have been given on p. 168f. as to why the synthesis of norsethite may serve as a model for dolomitization. An additional argument is the following. It is easy to prepare norsethite under suitable conditions, and it appears to be much more difficult to synthesize dolomite. In view of the many analogies existing between norsethite and dolomite, it is reasonable to conclude that the required conditions for the easy synthesis also represent the minimum requirements for the more difficult case of dolomite formation. It is felt, therefore, that the most successful path to dolomite at temperatures below 100° C should be analogous to reaction (24)

$$CaCO_{3\,(solid)} + Mg^{2+} + CO_3^{2-} \rightarrow CaMg(CO_3)_{2\,(solid)}. \qquad (29)$$

The present writer has attempted to prepare dolomite according to this equation under the prescribed conditions of norsethite formation. However, in the solutions specified on p. 173 no dolomite has so far been obtained, neither from calcite nor from aragonite as starting materials, at least not within periods of time on the order of a year. Otherwise much of the space allotted to the analogies with dolomite, to the synthesis of norsethite, and to its mechanism of formation could have been reserved for the description and interpretation of dolomite synthesis at room-temperature.

Success in this direction is still possible by using solutions higher in CO_3^{2-} than described on p. 173. In the case of dolomite, the congestion of the Mg^{2+} sites by the larger cation (Ca^{2+}) constitutes a more serious obstacle opposing crystal growth than in norsethite (Ba^{2+}). Here the sites of Mg and Ba are distinguished by different coordination numbers, 6 and 9 respectively. In dolomite both Ca and Mg sites have the same coordination number 6. Therefore, the chance that the more easily dehydratable large ions (Ca^{2+}) occupy Mg sites is still much greater in dolomite than for Ba^{2+} in norsethite. The situation would exemplify GOLDSMITH's (1953, see GRAF and GOLDSMITH, 1956) "simplexity principle" and may be one of the reasons for the metastable formation of magnesian calcites and disordered dolomites.

A correction for Ca^{2+} wrongly incorporated into Mg sites is conceivable only at low supersaturations, when the wayward Ca^{2+} has a chance to be released again into solution. This incorporation may be slight at extremely low concentrations of dissolved Ca^{2+} affected by higher CO_3^{2-} levels by virtue of the solubility product(s) of the calcium carbonate(s) present. In any case, both low supersaturation in general and a low level of dissolved Ca^{2+} in particular, as may be required for the formation of ordered dolomite, must be expected to slow down the crystallization rate to some unknown extent. It is nevertheless possible that the crystallization rate is fast enough for dolomite to form under experimental conditions within a period of time commensurate with the duration of a human life.

Even in that case, the reaction time will certainly be longer than that required for the formation of $PbMg(CO_3)_2$ from $PbCO_3$ as described on p. 171. Due to the slower rate, the circumstances of that reaction were more difficult to study even than those of norsethite formation and the results obtained for $PbMg(CO_3)_2$ were less clear-cut. From this point of view, the synthesis of norsethite will retain its importance as a model for dolomitization. If it should turn out that dolomite does not form at room-temperature within reasonable periods of time, then the syntheses of norsethite and $PbMg(CO_3)_2$ will remain our only sources of direct experimental information on the mode of formation of magnesium-bearing double carbonates under the conditions of the sedimentary cycle.

The first conclusion from the synthesis of norsethite is that sizable amounts of dolomite cannot precipitate directly from solution. So-called primary dolomites should, therefore, not exist. The calcium concentration in waters yielding dolomite should be at least as low as the barium concentration in norsethite formation. Deposits of calcareous sediments

12*

are, therefore, the only conceivable source of the calcium needed for any substantial quantity of dolomite to form, i.e. dolomite originates essentially by dolomitization of calcareous sediments as previously postulated by many investigators. It must be borne in mind nevertheless that the reaction proceeds completely via solution, since the calcium carbonate has to pass through the state of dissolved ionized species in order to react according to the true ionic reaction (16).

In most of the theoretical discussions on dolomitization published so far, reaction (15) or its specialized form (14) have been assumed as a quantitative description. Consequently, the main concern has been the supply of sufficient quantities of magnesium. In this connection the concept of seepage refluxion (ADAMS and RHODES, 1960) recently has become quite popular. The mechanism would take care also of the removal of the Ca^{2+} generated by reaction (15). This problem has been neglected in many speculations based on reaction (15), although it should be at least as important as the supply of magnesium. If, on the basis of our experience with norsethite, we accept the analogy to reaction (24),

$$CaCO_{3\,(solid)} + Mg^{2+} + CO_3^{2-} \rightarrow CaMg(CO_3)_{2\,(solid)}, \qquad (29)$$

as the most likely path to dolomite, we have to be concerned no longer about the supply of one ionic species and the removal of another. We need be concerned only with supply. This situation is simpler, offers less problems, and requires less complicated geological settings. In view of the low magnesium concentrations leading to norsethite and the ubiquity of that cation in marine environments, the formation of dolomite will depend largely upon the availability of CO_3^{2-}, i.e. alkalinity, which has to be supplied in stoichiometric amounts in addition to magnesium. Instead of attributing dolomitization to the coincidence of several geologic factors which must cooperate in other hypotheses, e.g. in seepage refluxion, the explanation of why calcium carbonate transforms to dolomite in one place and fails to do so elsewhere in a very similar environment, is now determined largely by one factor which is purely geochemical in character.

For dolomitization to proceed from sea water, all of the Ca^{2+} it contains must first be removed by precipitation as $CaCO_3$ (aragonite) before the process can even begin. Normal sea water contains 0.0104 M Ca^{2+} (Table 26) and 0.00146 M CO_3^{2-}, if the alkalinity of bicarbonate is calculated as carbonate. Thus, if CO_2 could be completely removed from sea water (and if there were no $MgCO_3^0$ complex), only 14% of the Ca^{2+} could precipitate as $CaCO_3$. An addition of at least 0.009 M CO_3^{2-} is thus necessary to depress the Ca^{2+} concentration to the level of the solubility of $CaCO_3$. Additional CO_3^{2-} is needed to promote reaction (29).

It is impossible that such radical changes in chemical composition occur in any bodies of sea water connected directly with the open sea. They can, however, persist in continental and lagoonal lakes and in the interstitial solutions of buried marine sediments.

We must now look for possible sources of the CO_3^{2-} required for the depression of the Ca^{2+} level and for reaction (29) to proceed in a stoichiometric fashion.

The origin of solutions containing alkali carbonate and bicarbonate as may occur on the continents has already been outlined in the chapter on alkali-bearing carbonate minerals (p. 91ff.). The Green River formation of Wyoming, which is famous for its bountiful occurrence of alkali-bearing carbonate minerals, contains considerable dolomite (FAHEY, 1962). It is at the same time the type locality of norsethite (MROSE et al., 1961). Both minerals could form there by means of the analogous reactions, (29) and (24), respectively, and the source of the CO_3^{2-} is obvious.

Dolomite in other lacustrine formations, e.g. in the European Keuper, may have formed as well in alkali-carbonate brines. Concentrations do not seem to have been sufficient for alkali-carbonate minerals to deposit, unless they have been removed by leaching during the course of the history of the formations in question.

On p. 93, the origin of the alkalinity in lacustrine water not balanced by calcium was traced back to the weathering of alkali- (and magnesium-) bearing silicate minerals, and the formation of carbonate and bicarbonate was ascribed to absorption of CO_2 from the air. Certain occurrences of dolomite in marine sediments (MURATA and ERD, 1964; PIERCE and MELSON, 1967; THOMPSON, BOWEN, MELSON and CIFELLI, 1968) and in near-shore environments (MÜLLER and TIETZ, 1966; ROTHE, 1968) can be directly correlated with the alteration of basalt and other basic rocks. In these occurrences, the magnesium set free by the alteration of the associated basic rocks is perhaps of accidental importance. Because of the availability of magnesium from sea water, it is hard to assess the relative importance of the two sources of Mg^{2+}. However, the alteration of the magnesian and alkaline constituents of the silicate rocks is a plausible source of alkalinity, i.e. of the CO_3^{2-} required to depress the Ca^{2+} level in the interstitial solution and to promote dolomite formation according to reaction (29).

For the lagoonal lakes of South Australia, VON DER BORCH (1965) has drawn attention to the seeping in of alkali-bicarbonate waters from the continent (p. 96). This influence may be responsible for the formation of recent dolomite in that locality. In Bonaire, Netherland's Antilles, the alteration of outcropping (and underlying?) volcanics and intrusives (DEFFEYES et al., 1965) may have been of greater importance in the formation of sub-recent dolomites than seepage refluxion.

A second source of alkalinity (or CO_3^{2-}) is anaerobic bacterial sulphate reduction. Its chief habitat is in buried sediments containing organic matter. It may also be active in the bottom waters of restricted sea basins and in lakes. The process has been known for quite some time (BEIJERINCK and VAN DELDEN, 1902; VAN DELDEN, 1903) and has been suggested as a mechanism leading to carbonate precipitation in certain marine environments (BAVENDAMM, 1932).

Chemical equations have been written which yield calcium carbonate as a product of the reduction of $CaSO_4$ or gypsum (e.g. ZOBELL, 1963; NEEV and EMERY, 1967) or which lead from sodium sulphate to bicarbonate (ABD-EL-MALEK and RIZK, 1963).

The stoichiometry of the bacterial reduction of the SO_4^{2-} and concomitant oxidation of organic matter can be described in a general fashion by the following two equations or linear combinations of both:

$$SO_4^{2-} + 2\{C\} + H_2O \rightarrow CO_3^{2-} + CO_2 + H_2S, \qquad (30)$$

$$SO_4^{2-} + 4\{H_2\} + CO_2 \rightarrow CO_3^{2-} + 3H_2O + H_2S. \qquad (31)$$

(These equations are in error in LIPPMANN, 1968b) Following a suggestion of I. A. Breger (kind personal communication), the C and H_2, serving as "food" for the sulphate-reducing bacteria and leading to the reduction of the sulphate, are written in braces. This is to denote carbon and hydrogen combined in appropriate organic compounds and to exclude the elements. Nevertheless, it appears that elemental hydrogen can be utilized by certain bacteria in the reduction of sulphate. However, since the hydrogen as such is not originally present in sediments, but is the product of bacterial decomposition of certain organic compounds (ZOBELL, 1947), the special case of elemental hydrogen may be left out of account.

Most organic compounds which are introduced into sediments by dead organisms (e.g. alcohols, carbohydrates, amino acids, fatty acids, and other organic acids), appear to be utilizable by sulphate-reducing bacteria. It would be beyond the scope of the present subject to list all organic compounds which are known to support the bacterial reduction of sulphate. A special survey of the pertinent literature would be required. The picture would nevertheless be incomplete, since it is doubtful whether all organic materials which may be incorporated into sediments have already been tested. However, it may reasonably be expected that a list of the organic compounds not utilizable by sulphate reducers would be much shorter.

What is important about sulphate reduction in relation to dolomite formation is the fact that sulphate-reducing bacteria and their spores are ubiquitous and that they become active as soon as they find an appropriate environment, i.e. anaerobic conditions and the presence of sulphate plus suitable organic matter. With these restrictions, sulphate-reducing bacteria are known to be viable over wide ranges of pH, temperature, salinity, and H_2S concentrations (see literature quoted by Zobell, 1963).

Since H_2S is an extremely weak acid, it will not greatly influence the equilibria related to dolomite formation. Therefore, its removal from the sediment is not essential for dolomitization to proceed. [The rôle of H_2S is different from that played by the Ca^{2+} liberated by the hypothetical reaction (15) which would act directly on the pertinent equilibrium.] It is nevertheless sound to assume that much of the H_2S produced will pass through the sediment-water interface by diffusion and as gas bubbles. Small amounts may be fixed in pyrite or remain in the pore space giving rise to fetid carbonate rocks.

Laboratory experiments have shown that sulphate-reducing bacteria continue their activity until all organic matter and/or available SO_4^{2-} are consumed. In buried sediments still in contact with sea water, SO_4^{2-} may be replenished by diffusion along the concentration gradient created by the consumption of SO_4^{2-}, or the supply may be secured by porous flow. The Mg^{2+} required for dolomite formation can be supplied by these mechanisms as well. In certain cases, the process of seepage refluxion as visualized by ADAMS and RHODES (1960) may have been active indeed in sustaining dolomitization by supplying both SO_4^{2-} and Mg^{2+}.

The amount of alkalinity produced by sulphate reduction will primarily depend upon the amount of organic matter initially contained in the sediment. It has been known since VAN DELDEN (1903) that the amounts of sulphate consumed and the alkalinity produced are equivalent stoichiometrically (see also ABD-EL-MALEK and RIZK, 1963). In a similar fashion, both these amounts must be equivalent to the quantity of organic matter consumed. Considering reaction (29), the amount of organic matter required to form a certain amount of dolomite can thus be calculated from Eqs. (30) and (31). The oxidation of 13.0 grams of organic carbon is equivalent to the formation of 100 grams of dolomite. To use a more realistic example, 32.6 grams of a carbohydrate $(CH_2O)_n$, when oxidized by SO_4^{2-}, will yield the alkalinity capable of forming 100 grams of dolomite according to reaction (29).

The simplest combination of reactions (30) and (31) would be the oxidation of methane:

$$SO_4^{2-} + CH_4 \rightarrow CO_3^{2-} + H_2O + H_2S. \qquad (30); (31)$$

Information in the literature as to whether methane is utilizable by sulphate-reducing bacteria is contradictory. 8.7 grams of CH_4 would be equivalent to 100 grams of dolomite. The reducing equivalent weight of hydrocarbons increases slightly with the number of carbons: 9.32 grams of ethane; 9.56 grams of propane, 9.70 grams of butane, to reach the limit of 10.14 grams for long chain hydrocarbons (C_nH_{2n+2}).

Dolomitization initiated in buried marine sediments by sulphate reduction starts essentially from the composition of sea water, although the concentrations in the pore space itself are radically changed with respect to the open sea. The CO_3^{2-} produced will at first give rise to the precipitation of the calcium dissolved in the pore solution as a carbonate, e.g. aragonite. Submarine cementation may, therefore, be regarded as a preliminary stage of dolomitization occurring most likely at a level in the sediment which is closer to the sediment-water interface than the site of dolomitization. It has been postulated above that the SO_4^{2-} consumed by sulphate reduction and also the Mg^{2+} to be fixed in dolomite are replenished by diffusion and/or porous flow. In the same way more Ca^{2+} will be transported to the site of sulphate reduction and calcium carbonate precipitation. In this connection it is important that sea water contains more SO_4^{2-} (0.0284 M) than Ca^{2+} (0.0104 M), so that alkalinity in excess of the precipitated calcium carbonate can be generated for dolomitization to proceed according to (29).

It is seen that a sizable portion, about one third, of the alkalinity produced by sulphate reduction, and the equivalent organic matter, are used up for the removal of the calcium from the interstitial solution. Therefore, the equivalent weights of organic materials quoted above will yield 100 grams of dolomite only when solutions devoid of dissolved Ca^{2+} are assumed. If sea water is considered as the reservoir solution and if SO_4^{2-} and Ca^{2+} are transported to the site of sulphate reduction and dolomitization at comparable rates, then the amounts of organic materials quoted above will yield about 60 to 70 grams of dolomite instead of 100 grams. Since Mg^{2+} is higher in sea water (0.0540 M) than SO_4^{2-} (0.0284 M), we need not be concerned about the supply of that constituent of dolomite.

The quantities of organic matter required for dolomitization according to (29) via (30) and (31) (and corrected for the precipitation of $CaCO_3$) do not appear excessive when compared to the amounts of ancient organic matter available as petroleum or contained in bituminous shales. In the latter, the oxidation by sulphate was incomplete, perhaps due to low permeability and ensuing insufficient supply of SO_4^{2-} from superjacent sea water.

That sulphate reduction may be an important geological process is attested to by H_2S contained in certain natural gases. The oil field of Lacq, Southern France, is a case in point. The gas produced there contains 11.3 vol% H_2S (STOJCIC, 1964). The reservoir rocks consist of porous carbonates including dolomite. If sulphate reduction did not affect the hydrocarbons themselves, it must have acted on their organic precursors. It is likely that the process was instrumental in the dolomitization of the carbonates of the Lacq reservoir.

BAVENDAMM (1932) noticed the production of abundant hydrogen sulphide in the recent calcareous sediments of the Bahamas, notably in mangrove swamps, and he drew attention to possible correlations between H_2S production and the precipitation of calcium carbonate. SHINN, LLOYD and GINSBURG (1969) report the formation of H_2S at or near localities on Andros Island, Bahama, which are known for the occurrence of recent dolomite, but they do not consider the possibility of an interdependence of the two phenomena.

Dolomites which owe their origin to alkalinity produced by sulphate reduction and concomitant oxidation of organic matter may be referred to as "organic" dolomites, in order to distinguish them from "inorganic" dolomites which have formed from alkalinity ultimately derived from the alteration of silicate minerals. Most dolomite formations have not so far been studied from this point of view.

Organic materials of biogenic origin are known to be deficient in the carbon isotope ^{13}C. Such a deficiency, or relative enrichment in the light carbon ^{12}C, should characterize dolomites which formed from alkalinity derived from sulphate reduction. The first example of "organic" dolomite containing isotopically light carbon was reported from the tar sandstone of Miocene age from Point Fermin, California, by SPOTTS and SILVERMAN (1966). The authors correlate the C-isotope composition of the dolomite with the (partial) oxidation of the associated bitumen and envisage sulphate ions among the possible oxidants.

Other examples of dolomite containing isotopically light carbon were reported by RUSSELL et al. (1967) from Recent marine sediments off the coast of Oregon; by MURATA, FRIEDMAN and MADSEN (1969) from the Monterey formation (Miocene) of California and Oregon; by DEUSER (1970) from the Quaternary of the continental shelf off the northeastern United States; and by TAN and HUDSON (1971) from the Great Estuarine Series of the Jurassic of Scotland. According to the latter authors, only part of the carbon of the dolomites can be derived from the same source as that of the coexisting calcites, which is isotopically heavier. The isotopically light carbon in the dolomites can be explained by the incorporation of biogenic carbonate through reaction (29).

The concentration of the light isotope in the biogenic carbon supplying alkalinity according to (30) must not be preserved in the final dolomite in every case. It is possible that the production of alkalinity from (30) and (31) is faster than dolomitization. In this situation, the alkalinity and the magnesium which later give rise to dolomitization may be fixed at first as hydrous magnesium carbonates (hydromagnesite and/or nesquehonite) precipitated in the pore space of the initially calcareous sediment. When sulphate reduction is no longer active, e.g. due to consumption of utilizible organic matter, a normal isotopic composition may be restored by porous flow for the carbonate species dissolved in the interstitial solution. In the temperature range of diagenesis, dolomite formation from preexisting solid calcium and hydrous magnesium carbonates is feasible only via an ionized solution according to (16). When the interstitial solution is further subject to porous flow, the isotopically light carbon of the hydrous magnesium carbonates may be homogenized with the carbonate species dissolved in the interstitial solution. In this manner, the original deficiency in ^{13}C stemming from sulphate reduction is obliterated by the final formation of dolomite in much the same way in which the high strontium contents are removed from originally aragonitic sediments in the course of the formation of calcitic limestones.

"Organic" dolomites, i.e. ancient formations which ultimately owe their origin to sulphate reduction and concomitant oxidation of organic matter are thus not necessarily characterized by a deficiency of ^{13}C, although the presence of isotopically light carbon appears to be a reliable indicator of that mechanism of formation. In the absence of light carbon, the question of from which source of alkalinity a given dolomite body has formed must be approached by considering all petrographic and geologic data available, and an answer can be expected only from the reconstruction of the history of the formation under study. In Table 27 a number of criteria are listed which may be helpful in distinguishing the two genetically different types of dolomite.

Evaporation and high salinity do not appear in Table 27 as factors conducive to dolomite formation. On p. 160f. it was shown that the supersaturation of sea water with regard to dolomite most probably is not increased by evaporation and high salinity. A direct interdependence of this factor and dolomite formation is thus very unlikely. The frequently observed and much discussed association of the two phenomena may be misleading. Evaporation can give rise to elevated salinity and to the deposition of salt minerals only in restricted or isolated basins. Any alkaline water discharged from the continent into such a basin will not be diluted as completely as it would in the open sea. In this manner, the alkalinity introduced into a limited body of sea water may lead to the removal of dissolved Ca^{2+} by the precipitation of $CaCO_3$ (aragonite). Any additional alkalinity may then give rise to dolomitization according to (29).

Restricted and isolated basins are at the same time preferred sites of sulphate reduction in bottom waters. Such basins may thus afford suitable environments for the formation of both "inorganic" and "organic" dolomites. From this point of view, the frequent occurrence of dolomite

Table 27. Criteria for the distinction of "organic" and "inorganic" dolomites

	Organic dolomites	Inorganic dolomites
Necessary condition	Reducing conditions, not necessarily in the water body, but at least in the buried sediment	
Sufficient, but not necessary	Isotopically light carbon	Association with alkali-bearing carbonate minerals or pseudomorphs of same. Oxidizing conditions (as e.g. in redbeds)
Characteristic environments. Most of these conditions are not cogent	Sediments of the open sea and of open lagoons Reefs, bioherms, biostromes Absence of sizable amounts of clastic and volcanic materials Absence of immediately overlying (and underlying) continental and lacustrine sediments	Continental and lacustrine Lagoons isolated from the open sea Contact with altered silicate rocks

in saline environments is a mere coincidence. In view of the possibility of the formation of "organic" dolomite in buried sediments of the open sea (RUSSEL et al., 1967; DEUSER, 1970), it is inappropriate to consider dolomites in ancient formations as indicators of elevated salinity.

The only possible connection between evaporation and dolomite formation is the formation of anhydrite. The solubility of gypsum is certainly too high to secure a sufficiently low concentration of dissolved Ca^{2+}. Although some information is available concerning the conditions of formation of anhydrite since the work of HARDIE (1967), our knowledge on the equilibrium solubility of the mineral is scanty. It is possible that to approach the solution of the "anhydrite problem" the same factors must be considered which complicate the formation of magnesite (see p. 82ff.). It may turn out that the true solubility product for anhydrite is lower than is now realized, say of the order of 10^{-8}. In that case, anhydrite might be instrumental in depressing the level of dissolved Ca^{2+}. In spite of the failure of USDOWSKI's (1967) experiments to yield dolomite in the sulphate system, it cannot be excluded that the Haidinger reaction (13), written for anhydrite, may lead to dolomite under certain conditions which are especially prone to favor the ready formation of anhydrite. Occurrences of dolomite which may have originated by the Haidinger reaction (13) should contain anhydrite at least in amounts stoichiometrically equivalent to the dolomite present. Hence, the mechanism could at most account for the origin of dolomite contained in anhydrite rocks, but not for dolomite containing admixtures of anhydrite or devoid of the mineral.

In the discussion of the formation mechanisms of "organic" dolomites we have considered the possibility of a preliminary fixation of Mg^{2+} and CO_3^{2-} ions as hydromagnesite and/or nesquehonite, which later react more slowly via solution with $CaCO_3$ to yield dolomite according to (29). Such preliminary reactions may be important also in the formation of "inorganic" dolomites. In point of fact, the hydromagnesite occurring in certain lagoonal lakes in South Australia is regarded by VON DER BORCH (1965) and ALDERMAN (1965) as a precursor of the dolomite and the magnesite which form in the same locality.

Another mineral which may be regarded as a precursor of dolomite and which has been found associated with recent dolomite on the Trucial Coast in the Persian Gulf is huntite (KINSMAN, 1967). We know very little of the conditions of formation of the mineral. Even so, huntite should form according to the overall reaction:

$$CaCO_{3\,(solid)} + 3\,Mg^{2+} + 3\,CO_3^{2-} \rightarrow CaMg_3(CO_3)_{4\,(solid)} \qquad (32)$$

if we generalize our experience with the synthesis of norsethite. For some reason or other, e.g. due to its more open structure (see p. 43 and p. 167), huntite may show a greater crystallization rate than dolomite, in spite of its greater solubility and of the greater number of Mg^{2+} ions which must be dehydrated. Thus huntite may serve as a reservoir of Mg^{2+} and CO_3^{2-} ions which may later react with admixed calcium carbonates to form dolomite.

Magnesium-deficient (or calcian) dolomites, which are usually characterized by cation disorder, have been assigned the name "protodolomite" by GRAF and GOLDSMITH (1956). The term implies that such phases act as precursors of well-ordered dolomites, not only in some of the experiments described by the authors, but also in nature. The latter view seems to have been confirmed by the fact that most of the dolomites found in Recent occurrences are indeed protodolomites according to the original definition. However, similar phases have been found to occur also in more ancient formations, e.g. in the Jurassic and in the Permian by FÜCHTBAUER and GOLDSCHMIDT (1965). "Protodolomites", therefore, may show considerable persistence in the course of geologic times.

There is so far no reliable evidence that calcian dolomites have indeed acted as precursors of (non-metamorphic) well-ordered dolomites.

For this reason the term protodolomite has not been used in other parts of this monograph.

The formation of "protodolomites" is usually ascribed to an environment deficient in magnesium. However, according to the analysis of dolomite formation presented in the first paragraphs of this chapter (p. 179), on the basis of the norsethite synthesis and the crystal structure of dolomite, the magnesium deficiency of certain dolomites may be due as well to insufficient availability of CO_3^{2-} ions in the original solution.

Many dolomite rocks are characterized by substantial porosities. These are usually greater than the porosities of associated limestones and of limestones in general. It has become customary to ascribe the porosity of dolomites, or at least part of it, to the change in volume which accompanies reaction (15). The volume of two moles (200.2 grams) of calcite (density 2.710), the solid starting material of reaction (15), is greater than the volume of one mole (184.4 grams) of dolomite (density 2.866). Thus, reaction (15) results in a decrease in solid volume of 12.9%, and non-porous calcite should yield dolomite characterized by this porosity value. Starting from calcite sediments of various initial porosities, the porosities of the resulting dolomite rocks may be calculated as well from the change in solid volume of 12.9%. It must, of course, be assumed than the situation is not complicated by compaction (WEYL, 1960). Even if dolomitization would proceed according to Eq. (15) in nature, as it does in the experiments of BUBB and PERRY (1968) at 200° C, a change in volume of 12.9% is by no means of fundamental importance. Aragonite as a starting material of dolomitization is as likely as calcite, and a change in solid volume of 5.8% is calculated for aragonite (density 2.930) when reacting according to (15). Therefore, the decrease in solid volume accompanying reaction (15) should rather be in the range between 5.8% and 12.9%, the exact value depending on the mineralogical composition of the original sediment.

Dolomitization according to reaction (29) is accompanied by a considerable increase in solid volume (see Table 28). The formation of porous dolomites is nevertheless conceivable provided that the process is completed under low overburden. The formation of porous dolomites must be a rather early process resulting in a solid framework which will resist compaction by any subsequent increase in overburden. The lower porosity of (associated) limestones then suggests that their lithification must have taken place under greater overburden and at a date much later than dolomitization.

Porosity in dolomites may also be interpreted as being a consequence of the preliminary formation of hydrous magnesium carbonates. It has been suggested above that hydromagnesite and/or nesquehonite may be precipitated as precursors of dolomite in the pore space of originally

Table 28. Changes in solid volume accompanying reactions assumed to yield dolomite

Type of reaction	Change in solid volume in % (starting volume = 100%; unless otherwise stated)	
	From calcite	From aragonite
(15)	-12.9	-5.81
(29)	$+74.2$	$+88.4$
(29) (final volume = 100%)	$+42.6$	$+46.9$
$Mg_4(CO_3)_4 \cdot Mg(OH)_2 \cdot 4H_2O^a + 5CaCO_3$	-19.0	-16.1
$MgCO_3 \cdot 3H_2O^b + CaCO_3$	-42.4	-40.9
$CaMg_3(CO)_3)_4{}^c + 2CaCO_3$	-1.86	$+0.98$

[a] Hydromagnesite, X-ray density = 2.199 (BALCONI and GIUSEPPETTI); the formula $Mg_3(CO_3)_3 \cdot Mg(OH)_2 \cdot 3H_2O$ in conjunction with a density of 2.15 as reported in the literature results in the same values for the change in solid volume as shown in this table.
[b] Nesquehonite, density = 1.85.
[c] Huntite, X-ray density = 2.874.

calcareous sediments. The formation of dolomite from such mixtures at a later date would yield porosities from 16% to over 40% (see Table 28). This range of porosities is not at variance with values actually determined for dolomite rocks (cf. FÜCHTBAUER and GOLDSCHMIDT, 1965). The final porosities are, of course, subject also to the influence of initial porosities and compaction.

It is worth noting that huntite (density 2.874) assumed as a precursor of dolomite, would yield rather slight changes in solid volume (Table 28) when reacting with the appropriate amount of $CaCO_3$.

E. Petrological Summary: Reaction Series Leading from Carbonate Sediments to Carbonate Rocks

I. The Formation of Fresh-Water Limestones

The simplest process of carbonate sedimentation is the precipitation from fresh-water, i.e. from solutions which are so low in dissolved magnesium that calcium is essentially the only important cation susceptible to forming a sparingly soluble carbonate. One of the most important factors determining carbonate solubility in any natural water is the partial pressure of CO_2, which acts upon the activity product of the carbonate to precipitate via the equilibria (9), (10), and (11) (p. 103) among the various dissolved species related to the carbonate ion.

This interplay is demonstrated in a particularly clear fashion in many localities of recent carbonate precipitation from fresh water where the process can be correlated with the escape of CO_2 into the atmosphere or with the consumption of CO_2 by the photosynthetic activity of aquatic plants. These types of interdependence show that fresh water does not sustain extreme supersaturations with respect to calcium carbonate. In most cases, the phase precipitated is calcite, the stable polymorph of calcium carbonate characterized by the lowest solubility product. In general, supersaturation in fresh waters is not capable of exceeding the critical ionic product for aragonite.

Calcite sediments forming from fresh water are often more (travertine) or less (calcareous tufa) lithified as they precipitate. In the light of results obtained from experimental carbonate precipitations, this feature and the predominantly calcitic composition of fresh-water carbonates may be ascribed to the scarcity of dissolved magnesium in such environments.

Fresh-water limestones are of limited quantitative importance. Most known occurrences are of Quaternary age, and well-documented examples in more ancient formations are rare. They are nevertheless extremely instructive in that equilibrium conditions are more closely obeyed than in marine carbonate sedimentation. Fresh-water limestones are also remarkable for the relatively short times during which consolidated carbonate rocks may form. The process may be so straightforward under near-equilibrium conditions that it is even an exaggeration to speak of a reaction series relating to fresh-water limestone formation.

In unconsolidated calcitic muds, interstitial precipitation and partial dissolution followed by reprecipitation are the most likely mechanisms leading to solid limestones. Under low overburden dissolution and reprecipitation may be brought about by changes in CO_2 pressure, whereas pressure solution can become active at greater depths of burial.

II. The Evolution of Marine Limestones

Carbonate solubility in sea water is governed essentially by the same parameters as in fresh water. However, the activities entering into the solubility products and into the equilibria [(9), (10), (11), p. 103], which interrelate the partial pressure of CO_2 with the activity of the carbonate ion, must be determined to comply with the presence of additional dissolved species and account must be taken of complex formation (see Table 26). The precipitation of calcium carbonate is complicated by the specific influence of dissolved magnesium. Judging from what is known about the solubility products of dolomite and magnesite, the surface waters of warm seas must be considerably supersaturated with respect to these carbonates. They should be the products of marine carbonate precipitation, if equilibration were spontaneous. However, due to the dehydration barrier, the incorporation of magnesium into anhydrous carbonates is strongly inhibited at earth-surface temperatures.

Since aragonite is not disposed to incorporate sizable amounts of magnesium into its structure, it is the only sparingly soluble carbonate which precipitates freely from sea water as soon as a sufficient supersaturation prevails. The most important factor seems to be temperature, which promotes the escape of CO_2 into the atmosphere. The degree of supersaturation required for aragonite growth depends to a great extent upon the availability of nuclei. Supersaturations giving rise to spontaneous nucleation, e.g. in the form of whitings, seem to be restricted to the warmest parts of the seas.

Amounts of the order of 1% strontium are passively incorporated into marine aragonites.

The conditions for the inorganic formation of magnesian calcites from sea water are not well understood. Their occurrence as a cement in the pore space of marine carbonate sediments soaking in sea water shows that they may form under conditions similar to those yielding aragonite. It is suggested that magnesian calcites require lower supersaturations to grow than aragonite, but the growth rate is much slower due to the incorporation of magnesium. Thus appreciable amounts of magnesian calcites seem to crystallize preferably within the sediment where physicochemical conditions remain constant for more extended periods of time than is the case in shallow warm seas, the chief habitat for the inorganic formation of aragonite.

Both aragonite and magnesian calcites are metastable phases which ultimately owe their origin to the reluctance of the magnesium ion to be dehydrated and to be incorporated into the stable structures of dolomite and magnesite.

Although the situation may be modified to some extent by organic processes and by the presence of organic compounds, the metastable carbonates may be assumed to form in the hard parts of marine organisms for largely the same reasons as in the sea. In the form of organic debris, not only aragonite and magnesian calcites are introduced into sediments, but also essentially pure calcite. This mineral can be expected to segregate from body fluids which are lower in dissolved magnesium than sea water. It appears that the organic membranes controlling the interchange between the body fluid and the open sea can be more or less ion-selective in various organisms. On the whole, the mechanisms responsible for the segregation of the various forms of calcium carbonate in organisms are far from being completely understood. Nevertheless, in view of the high Mg^{2+} level in sea water, organic hard parts are the only source for primary, essentially pure calcite in marine sediments. This type of calcite may resist many of the diagenetic changes to which a sediment is later subjected.

The metastable carbonates, aragonite and magnesian calcites, will persist in a sediment as long as its pores are filled by essentially unmodified sea water. Carbonate precipitation in the pore space may lead to submarine cementation by aragonite and/or magnesian calcites. The most effective driving force for such precipitation within the sediment appears to be bacterial sulphate reduction. In the absence of this type of preliminary lithification, marine carbonate sediments soaking in sea water may remain friable for millions of years.

For the metastable carbonates to convert to stable calcite, the magnesium dissolved in the pore solution must be removed. The most effective mechanism is the replacement of the original pore solution by fresh water. To this end, a calcareous sediment must normally be removed from the marine environment by way of tectonic movements or sea-level fluctuations.

In a pore solution free from dissolved magnesium, the kinetic obstacles which so far prevented the crystallization of rhombohedral carbonates no longer exist. Any preexisting or newly forming nuclei of calcite have now free rein to grow at the expense of more soluble aragonite and magnesian calcites. This recrystallization via solution is the most important step on the way from loose or cemented calcareous sediments to hard limestones.

13 Lippmann

Table 29. Outline of processes active in the formation of marine carbonate rocks

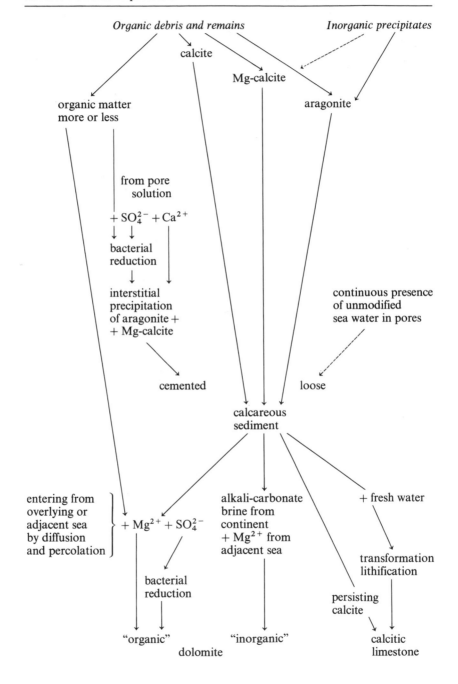

Sediments devoid of recrystallizable metastable carbonates, i.e. purely calcitic organic debris may be expected to remain friable for long periods of time, even after the pore space has been invaded by fresh water.

The evolution of marine limestones as summarized above is outlined in Table 29 along with the processes of dolomitization to be summarized below.

III. Dolomitization

Sea water is considerably supersaturated with respect to dolomite. Nevertheless, the mineral does not precipitate from normal sea water or from more concentrated brines. This is due to kinetic inhibitions arising from the strength of the $H_2O—Mg^{2+}$ bond and the low CO_3^{2-} activity. For the same reasons, sea water and related solutions are incapable also of reacting with calcareous sediments to form dolomite, although the reaction should proceed in this direction according to equilibrium thermodynamics.

Reaction of calcareous sediments with magnesium-bearing solutions is, nevertheless, the only way conducive to forming sizable amounts of dolomite. However, the reacting solutions must be virtually free from dissolved calcium. The ion must be eliminated from solution by precipitation as a carbonate. To this end, dissolved CO_3^{2-} in stoichiometric excess over the calcium dissolved in sea water must be supplied. Equivalent amounts of CO_3^{2-} and Mg^{2+} are then consumed by dolomitization proceeding at the expense of preexisting calcareous sediment. The overall reaction is:

$$CaCO_{3\,(solid)} + Mg^{2+} + CO_3^{2-} \rightarrow CaMg(CO_3)_{2\,(solid)}. \qquad (29)$$

In view of the ubiquity of magnesium in marine and related environments, the question of whether a calcareous sediment ends up as a limestone or as a dolomite rock is largely decided by the availability of CO_3^{2-} or, more generally speaking, alkalinity.

In continental and near-shore environments, alkaline waters, which ultimately owe their origin to the alteration of silicates, may supply the alkalinity required for dolomitization.

An alternative source of alkalinity is bacterial sulphate reduction, which may be active in the deeper parts of isolated water bodies and in buried sediments. Sulphate reduction is sustained by the concomitant oxidation of organic matter. Dolomites formed at the expense of alkalinity generated by sulphate reduction may, therefore, be referred to as "organic" dolomites, in order to distinguish them from "inorganic" dolomites which have originated from alkalinity set free by inorganic processes, such as silicate decomposition.

13*

Appendix:
Aids to the Determination of Sedimentary Carbonate Minerals

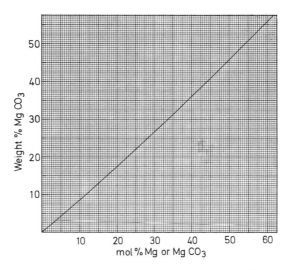

Fig. 54. Relationship between weight and mole percentage in magnesian calcites.
Note that the graph is not a straight line, but a slightly though distinctly bent curve

Tables of X-ray diffraction powder data for some carbonate minerals and brucite.

Table 30. X-ray powder data for simple rhombohedral carbonates (d in Å)

hkil (hex)	Calcite CaCO$_3$ d	Int.	Rhodochrosite MnCO$_3$ d	Int.	Siderite FeCO$_3$ d	Int.	Magnesite MgCO$_3$ d	Int.	hkl (rh)
01$\bar{1}$2	3.8551	12	3.6581	35	3.5903	25	3.5387	—	110
10$\bar{1}$4	3.0359	100	2.8440	100	2.7912	100	2.7412	100	211
0006	2.8440	3	2.6107	—	2.5622	—	2.5027	17	222
11$\bar{2}$0	2.4949	14	2.3886	20	2.3443	20	2.3165	4	10$\bar{1}$
11$\bar{2}$3	2.2848	18	2.1721	27	2.1318	27	2.1023	43	210
20$\bar{2}$2	2.0946	18	2.0000	23	1.9629	30	1.9382	12	002
02$\bar{2}$4	1.9275	5	1.8290	12	1.7952	15	1.7693	3	220
01$\bar{1}$8	1.9127	17	1.7698	30	1.7369	35	1.7002 } 34		332
11$\bar{2}$6	1.8755	17	1.7623	33	1.7296	44	1.7000 }		321
21$\bar{3}$1	1.6259	4	1.5559	1	1.5271	—	1.5088	4	20$\bar{1}$
12$\bar{3}$2	1.6042	8	1.5334	13	1.5050	19	1.4865	5	21$\bar{1}$
10$\bar{1}$10	1.5872	2	1.4649	—	1.4377	—	1.4063 } 4		433
21$\bar{3}$4	1.5253	5	1.4522	1	1.4253	16	1.4061 }		310
20$\bar{2}$8	1.5180	4	1.4220	1	1.3956	7	1.3706	3	422
11$\bar{2}$9	1.5096	3	1.4066	—	1.3805	—	1.3538 } 7		432
12$\bar{3}$5	1.4733	2	1.3991	—	1.3732	—	1.3537 }		320
03$\bar{3}$0	1.4404	5	1.3790	10	1.3535	20	1.3374	8	11$\bar{2}$
00012	1.4220	3	1.3053	<1	1.2811	—	1.2513	3	444
21$\bar{3}$7	1.3569	1	1.2817	—	1.2580	—	1.2383	<1	421
02$\bar{2}$10	1.3391	2	1.2488	<1	1.2256	—	1.2021	<1	442
12$\bar{3}$8	1.2968	2	1.2219 }		1.1992 }		1.1796 }		431
30$\bar{3}$6 }				3		17		<1	411
03$\bar{3}$6 }	1.2850	1	1.2194 }		1.1968 }		1.1796 }		330
22$\bar{4}$0	1.2475	1	1.1943	—	1.1722	—	1.1583	<1	20$\bar{2}$
22$\bar{4}$3	1.2185	—	1.1642	—	1.1427	—	1.1284	<1	31$\bar{1}$
11$\bar{2}$12	1.2354	2	1.1454 } 1		1.1242	—	1.1010	<1	543
13$\bar{4}$1	1.1956	—	1.1444 }		1.1232	—	1.1098	—	21$\bar{2}$
31$\bar{4}$2	1.1869	—	1.1353	—	1.1143	—	1.1008	(<1)	30$\bar{1}$
21$\bar{3}$10	1.1799	3	1.1066 } 1		1.0861	17	1.0670 } 4		532
13$\bar{4}$4	1.1539	3	1.1011 }		1.0807	26	1.0669 }		32$\bar{1}$
01$\bar{1}$14	1.1731	(3)	1.0801	—	1.0600 } 17		1.0362	—	554
22$\bar{4}$6	1.1424	1	1.0860	?	1.0659 }		1.0511	1	420

The d-spacings are from GRAF (1961). They were calculated by GRAF from the cell dimensions quoted in Table 3.

The intensities for calcite, rhodochrosite, and magnesite are reduced diffractometer readings determined by SWANSON et al. (1953, 1957). The intensities for siderite are from SHARP, W.E., Am. Mineralogist **45**, 241—243 (1960).

Table 31. X-ray powder data for rhombohedral double carbonates (d in Å)

hk.l (hex)	Dolomite $CaMg(CO_3)_2$ d	Int.	Ankerite $Ca(Mg, Fe)(CO_3)_2$ d	Int.	$SrMg(CO_3)_2$ d	Int.	Norsethite $BaMg(CO_3)_2$ d	Int.	hkl (rh)
00.3	5.3366	—	5.367	?	5.480	6	5.590	23	111
10.1	4.0297	3	4.040	—	4.113	14	4.196	44	100
01.2	3.6939	5	3.704	3	3.774	16	3.858	45	110
10.4	2.8855	100	2.899	100	2.954	100	3.017	100	211
00.6	2.6683	10	2.685	3	2.740	—	2.795	1	222
01.5	2.5382	8	2.552	2	2.600	24	2.655	29	221
11.0	2.4039	10	2.411	3	2.453	20	2.509	29	10$\bar{1}$
11.3	2.1918	30	2.199	6	2.239	26	2.289	21	210
02.1	2.0645	5	2.067	1	2.106	16	2.154	23	11$\bar{1}$
20.2	2.0149 } 15		} 2.020	3	2.056 } 33		2.103 } 26		{ 200
10.7	2.0046				2.055		2.098		{ 322
02.4	1.8470	5	1.852	1	1.887	9	1.929	7	220
01.8	1.8037	20	1.812	6	1.850	31	1.888	18	332
11.6	1.7860 } 30		} 1.792	6	1.827 } 42		1.867 } 22		{ 321
00.9	1.7789				1.827		1.863		{ 333
20.5	1.7454	—	1.751	—	1.784	—	1.823	1	311
21.1	1.5662	8	1.569	1	1.598	6	1.634	3	20$\bar{1}$
12.2	1.5442 } 10		} 1.548	2	1.576 } 14		1.612 } 12		21$\bar{1}$
02.7	1.5396				1.575		1.609		331
10.10	1.4943	2	1.501	<1	1.533	4	1.564	3	433
21.4	1.4646	5	1.468	2	1.496	12	1.529	11	310
20.8	1.4428	4	1.449	3	1.477	8	1.508	5	422
11.9	1.4300	10	1.436	1	1.465	5	1.496	4	432
12.5	1.4124	4	1.416	<1	1.443	6	1.475	6	320
03.0	1.3879	15	1.391	1	1.416	6	1.448	4	11$\bar{2}$
01.11	1.3739	—	1.381	—	1.410	—	1.439	1	443
30.3 } 03.3	1.3432	—	1.347	—	1.371 } 6		1.402	3	{ 300 22$\bar{1}$
00.12	1.3342	8	1.341	1	1.370		1.397	1	444
21.7	1.2965	4	1.300	<1	1.325	5	1.354	3	421
02.10	1.2691	4	1.273	<1	1.300	6	1.327	3	442
12.8	1.2371 } 5		} 1.241	1	1.265	5	1.293	4	431
03.6 } 30.6	1.2313 }				} 1.258	3	} 1.286	3	{ 41$\bar{1}$ 330
22.0	1.2020	3	1.205	<1	1.226	4	1.254	4	20$\bar{2}$

The d-spacings for dolomite are those calculated by GRAF (1961) from the cell dimensions of the ideally ordered mineral quoted in Table 4.

The intensities given for dolomite and the data for the ankerite, $Ca_{1.03}Mg_{0.63}Fe_{0.33}(CO_3)_2$, are observed values from HOWIE and BROADHURST (1958) (cf. Table 4).

The d-spacings for $SrMg(CO_3)_2$ (FROESE, 1967) and norsethite are calculated values which are based on the cell dimensions quoted in Table 4.

Table 32. X-ray powder data for huntite, $CaMg_3(CO_3)_4$

hk.l (hex)	hkl (rh)	d in Å	Int.	hk.l (hex)	hkl (rh)	d in Å	Int.
10.1	100	5.670	3	32.1	$30\bar{2}$	1.836	5
11.0	$10\bar{1}$	4.753	2	04.2	$22\bar{2}$	1.821	5
02.1	$11\bar{1}$	3.643	5	14.0	$21\bar{3}$	1.797	5
01.2	110	3.532	*	02.4	220	1.766	30
21.1	$20\bar{1}$	2.891	50	22.3	$31\bar{1}$	1.756	31
20.2	200	2.835	100	23.2	$31\bar{2}$	1.701	6
03.0	$11\bar{2}$	2.744	4	21.4	310	1.656	*
00.3	111	2.607	20	05.1	$22\bar{3}$	1.611	*
12.2	$21\bar{1}$	2.435	18	33.0	$30\bar{3}$	1.584	16
22.0	$20\bar{2}$	2.377	18	01.5	221	1.537	*
11.3	210	2.286	10	24.1	$31\bar{3}$	1.526	4
13.1	$21\bar{2}$	2.192	14	50.2	$11\bar{4}$	1.518	*
40.1	$\bar{3}\bar{1}1$	1.990	30	13.4	$32\bar{1}$	1.485	} 8
31.2	$30\bar{1}$	1.972	50	14.3	$32\bar{2}$	} 1.479	
10.4	211	1.902	} 2	41.3	$40\bar{1}$		
30.3	300	} 1.890					
03.3	$22\bar{1}$						

The d-spacings are from GRAF and BRADLEY (1962). They were calculated by the authors on the basis of the lattice constants quoted in Table 10. The (relative) intensities are from KINSMAN (1967) who did not observe the reflections marked by asterisk. These were noticed by GRAF and BRADLEY to be present with faint intensities.

Table 33. X-ray powder data for aragonite-type carbonate minerals (d in Å)

hkl	Aragonite $CaCO_3$ d	Int.	Strontianite $SrCO_3$ d	Int.	Witherite $BaCO_3$ d	Int.
110	4.212	2	4.367	14	4.56	9
020	3.984	—	4.207	6	4.45	4
111	3.396	100	3.535	100	3.72	100
021	3.273	52	3.450	70	3.68	53
002	2.871	4	3.014	22	3.215	15
121	2.730	9	2.859	5	3.014	—
012	2.700	46	2.838	20	3.025	4
102	2.489 ⎫	33	2.596	12	2.749	3
200	2.481 ⎭		2.554	23	2.656	11
031	2.409	14	2.543	—	2.695	—
112	2.372	38	2.481	34	2.628	24
130	2.341	31	2.458	40	2.590	23
022	2.328	6	2.4511	33	2.606	—
211	2.188	11	2.2646	5	2.367	—
220	2.106	23	2.1831	16	2.281	6
040	1.992	—	2.1035	7	2.226	2
221	1.977	65	2.0526	50	2.150	28
041	1.882	32	1.9860	26	2.104	12
202	1.877	25	1.9489	21	2.048	10
132	1.814	23	1.9053	35	2.019	21
141	1.759	4	1.8514	3	1.956	—
113	1.742	25	1.8253	31	1.940	15
023	1.725 ⎫	15	1.8134	16	1.931	—
231	1.728 ⎭		1.8023	4	1.892	—
222	1.698	3	1.7685	7	1.859	3
042	1.636	—	1.7253	5	1.830	2
310	1.618	—	1.6684	3	⎫ 1.737	2
033	⎫ 1.553	—	1.6335	—	⎭	
240	⎭		1.6236	4	1.706	1
311	1.557	4	1.6080	13	1.677	5

Compiled from SWANSON et al. (1953, 1954). The lattice constants which the authors used in indexing these data are contained in Table 18.

Table 34. X-ray powder data for vaterite from H.J.MEYER (1969)

hk.l	d in Å	Int.	hk.l	d in Å	Int.
00.4	4.23	21	41.0	1.353	7
11.0	3.57	57	41.2	1.334	7
11.2	3.30	100	11.12	1.313	13
11.4 ⎫			41.4	1.287	15
20.3 ⎭	2.73	94	33.0	1.192	2
21.1	2.32	6	33.2	1.181	1
20.5	2.29	2	30.12	1.167	5
11.6	2.22	6	33.4	1.148	10
21.3	2.16	4	41.8	1.140	12
00.8	2.117	15	22.12	1.108	7
30.0	2.065	62	00.16	1.056	3
30.4	1.858	26	33.8	1.035	1
11.8	1.823	72	60.0	1.032	1
22.0	1.788	4	11.16	1.018	10
22.2	1.749	1	30.16	0.942	6
?	1.707	1			
22.4 ⎫					
31.3 ⎭	1.647	26			
40.1	1.545	5			
31.5	1.511	1			
30.8	1.478	7			
00.12	1.416	1			
22.8	1.366	8			

The observed d-spacings were indexed by MEYER on the basis of the cell dimensions $a_{hex} = 7.151$ Å and $c_{hex} = 16.94$ Å. The indexing in terms of the cell shown in Fig. 29 is obtained by halving the l indices.

A number of l indices would then be fractional. Therefore the description by the cell characterized by $c' = \frac{c}{2} = 8.47$ Å is only an approximation.

Table 35. X-ray powder data for brucite, $Mg(OH)_2$

hk.l	d in Å	Int.	hk.l	d in Å	Int.
00.1	4.77	90	20.1	1.310	11
10.0	2.725	6	00.4	1.192	2
10.1	2.365	100	20.2	1.183	9
10.2	1.794	56	11.3	1.118	1
11.0	1.573	36	10.4	1.092	3
11.1	1.494	18	20.3	1.034	5
10.3	1.373	16	21.0	1.030	1
20.0	1.363	2			

From SWANSON,H.E., GILFRICH,N.T., and COOK,M.I., 1956, Standard X-ray diffraction powder patterns: U.S. Natl. Bur. Standards Circ. **539**, VI.

The observed d-spacings were indexed by the authors on the basis of the cell dimensions $a_{hex} = 3.147$ Å and $c_{hex} = 4.769$ Å.

Table 36. X-ray powder data for hydromagnesite (see p. 71f.)

hkl	d in Å	Int.	hkl	d in Å	Int.
200	9.30	5	802 613	2.02	0.5
210	6.46	5			
111	5.84	10	304 042 730 812	1.99	3
400	4.60	0.5			
020	4.49	2			
002	4.22	3			
220	4.07	1	441	1.97	0.5
012	3.84	2	731 540 703	1.93	2
500 411	3.70	0.5			
212	3.53	2	404 024	1.91	1
321	3.33	3			
420	3.24	0.5	10 0 0 822 902	1.85	1
511	3.18	0.5			
600	3.10	0.5			
610 222	2.92	10	10 10 912	1.81	1
131 003	2.80	1	604 251 350	1.74	2
013	2.69	4			
602	2.50	4	10 12 931	1.67	0.5
612	2.40	1			
530	2.34	2	714 840 434	1.62	4
800	2.32	3			
721 240	2.20	2	405 10 30	1.58	0.5
630	2.157	6	534 452 650	1.56	2
004	2.10	0.5			

From BALCONI and GIUSEPPETTI (1959).

The observed d-spacings were indexed by the authors on the basis of the cell: $a = 18.54$ Å, $b = 9.05$ Å, $c = 8.42$ Å, $\alpha = 90°$.

Table 37. X-ray powder data for artificial nesquehonite, $MgCO_3 \cdot 3H_2O$

d in Å	Int.	d in Å	Int.
7.25	4	2.175	29
6.61	94	2.165	10
6.46	88	2.135	4
6.07	12	2.125	5
4.93	8	a number of	
4.14	6	very faint	
3.88	100	reflections	
3.85	96	2.023	34
3.59	63	2.010	5
3.24	45	1.941	3
3.14	3	1.927	80
3.04	92	1.914	4
2.99	10	1.893	10
2.91	3	1.855	5
2.89	2	1.848	10
2.79	37	1.838	10
2.63	45	1.800	18
2.515	85	1.774	4
2.475	4	1.728	6
2.40	3	1.719	10
2.34	10	1.697	4
2.315	8	1.684	3
a number of		1.661	3
very faint		1.650	18
reflections		1.634	6
2.205	4	1.618	5
2.185	6		

LIPPMANN, unpublished. The doubling of the reflections at 6.46 Å and 3.85 Å does not occur in all cases. No reasons for its occurrence or absence can be given so far.

Appendix

Table 38. X-ray powder data for artificial monohydrocalcite, $CaCO_3 \cdot H_2O$

hk.l	d in Å	Int.	h'k'.l	hk.l	d in Å	Int.	h'k'.l
11.0	5.275	25	10.0	30.3	1.941	12	11.3
11.1	4.327	100	10.1	41.1	1.928	25	21.1
11.2	3.068	50	10.2	22.3	1.820	6	20.3
30.0	3.046	3	11.0	31.3	1.784 ⎱	6	—
30.1	2.823	25	11.1	11.4	1.775 ⎰		10.4
00.3	2.513	3	00.3	41.2	1.763	25	21.2
22.1	2.493	6	20.1	33.0	1.757	6	30.0
30.2	2.370	25	11.2	33.1	1.713	3	30.1
11.3	2.270	6	10.3	30.4	1.603	1	11.4
22.2	2.161	25	20.2	33.2	1.595	3	30.2
41.0	1.994	6	21.0	41.3	1.562	3	21.3

The *d*-spacings were calculated from the cell dimensions $a_{hex} = 10.553$ Å and $c_{hex} = 7.54$ Å as derived from single-crystal photographs (LIPPMANN, 1959). The value for $a_{hex} = 10.62$ Å in the original publication has been refined by calibration with NaCl.

The reflections *hk.l* not satisfying the condition $h^2 + k^2 + hk = 3n$ are weak in single-crystal photographs and they appear to be missing in X-ray diffractometer traces. Hence, the powder pattern may also be indexed in terms of a pseudocell with $a'_{hex} = a_{hex}/\sqrt{3} = 6.093$ Å by the indices *h'k'.l*.

References

ABD-EL-MALEK, Y., RIZK, S. G.: Bacterial sulphate reduction and the development of alkalinity. J. appl. Bacteriol. **26**, 7—26 (1963).

ADAMS, J. E., RHODES, M. L.: Dolomitization by seepage refluxion. Bull. Am. Ass. Petroleum Geol. **44**, 1912—1920 (1960).

ADAMSON, A. W.: The physical chemistry of surfaces, 187—188. New York-London: Interscience 1960.

ALDERMAN, A. R.: Dolomitic sediments and their environment in the South-East of South Australia. Geochim. Cosmochim. Acta **29**, 1355—1365 (1965).

ALDERMAN, A. R., SKINNER, H. C. W.: Dolomite sedimentation in the South-East of South Australia. Am. J. Sci. **255** 561—567 (1957).

ALÉONARD, S., VICAT, J.: Borates de structure dolomie. Bull. Soc. Franç. Minéral. Cristal. **89**, 271—272 (1966).

ANDALIB, F.: Mineralogisch-geochemische Untersuchungen der aragonitischen Fossilien aus dem Dogger alpha (Opalinuston) in Württemberg. Arb. Geol.-Paläont. Inst. Univers. Stuttgart Nr. 62, 1970.

ANDON, R. J. L., COX, J. D.: Phase relationships in the pyridine series Part I. The miscibility of some pyridine homologues with water. J. chem. Soc. (Lond.) **1952**, 4601—4606 (1952).

APPLEMAN, D. E.: X-ray crystallography of wegscheiderite, $Na_2CO_3 \cdot 3\ NaHCO_3$. Am. Mineralogist **48**, 404—406 (1963).

BÄCKSTRÖM, H. L. J.: The thermodynamic properties of calcite and aragonite. J. Amer. chem. Soc. **47**, 2432—2442 (1925).

BALCONI, M., GIUSEPPETTI, G.: Sull' idromagnesite della grotta De su Marmori (Sardegna). Studi Ric. Ist. Min. Petr. Univ. Pavia **1**, 1—24 (1959).

BALL, M. M.: Carbonate sand bodies of Florida and the Bahamas. J. Sediment. Petrol. **37**, 556—591 (1967).

BALLMAN, A. A.: A new series of synthetic borates isostructural with the carbonate mineral huntite. Am. Mineralogist **47**, 1380—1383 (1962).

BARON, G.: Précipitation de la giobertite et de la dolomie à partir des solutions de chlorures de magnésium et de calcium. Compt. Rend. **247**, 1606—1608 (1958).

BARON, G., PESNEAU, M.: Sur existence et un mode de préparation du monohydrate de carbonate de calcium. Compt. Rend. **243**, 1217—1219 (1956).

BATHURST, R. G. C.: Boring algae, micrite envelopes and lithification of molluscan biosparites. Geol. J. **5**, 15—32 (1966).

BATHURST, R. G. C.: Oölitic films on low energy carbonate sand grains, Bimini Lagoon, Bahamas. Marine Geol. **5**, 89—109 (1967a).

BATHURST, R. G. C.: Depth indicators in sedimentary carbonates. Marine Geol. **5**, 447—471 (1967b).

BATHURST, R. G. C.: Precipitation of oöids and other aragonite fabrics in warm seas. In: MÜLLER, FRIEDMAN (Eds.): Recent developments in carbonate sedimentology in Central Europe, pp. 1—10. Berlin-Heidelberg-New York: Springer 1968.

BATHURST, R. G. C.: Carbonate sediments and their diagenesis. Amsterdam—London—New York: Elsevier 1971.

BAUR, W.: Die Kristallstrukturen von $Mg_2SO_4 \cdot 4H_2O$ (Leonhardtit) und $FeSO_4 \cdot 4H_2O$ (Rozenite). Acta Cryst. **15**, 815—826 (1962).

BAUR, W.: Refinement of the crystal structure of $MgSO_4 \cdot 7H_2O$ (epsomite). Acta Cryst. **17**, 1361—1369 (1964).

BAUSCH, W. M.: Dedolomitisierung und Recalcitisierung in fränkischen Malmkalken. Neues Jahrb. Mineral. Monatsh. **1965**, 75—82.

BAUSCH, W. M.: Outlines of distribution of strontium in limestones. In: MÜLLER, FRIEDMAN (Eds.): Recent developments in carbonate sedimentology in Central Europe, pp. 106—115. Berlin-Heidelberg-New York: Springer 1968.

BAVENDAMM, W.: Die mikrobiologische Kalkfällung in der tropischen See. Arch. Mikrobiol. **3**, 205—276 (1932).

BEIJERINCK, M. W., VAN DELDEN, A.: Über die Assimilation des freien Stickstoff durch Bakterien. Centralbl. Bakt. (Abt. 2) **9**, 2—43 (1902).

BERNER, R. A.: Comparative dissolution characteristics of carbonate minerals in the presence and absence of aqueous magnesium ion. Amer. J. Sci. **265**, 45—70 (1967).

BISCHOFF, J. L.: Kinetics of calcite nucleation: Magnesium ion inhibition and ionic strength catalysis. J. Geophys. Res. **73**, 3315—3322 (1968).

BISCHOFF, J. L.: Temperature controls on aragonite-calcite transformation in aqueous solution. Am. Mineralogist **54**, 149—155 (1969).

BISCHOFF, J. L., FYFE, W. S.: The aragonite-calcite transformation. Am. J. Sci. **266**, 65—79 (1968).

BLACKMON, P. D., TODD, R.: Mineralogy of some foraminifera as related to their classification and ecology. J. Paleontol. **33**, 1—15 (1959).

BOETTCHER, A. L., WYLLIE, P. J.: The calcite-aragonite transition measured in the system $CaO\text{-}CO_2\text{-}H_2O$. J. Geol. **76**, 314—330 (1968).

BØGGILD, O. B.: The shell structure of molluscs. Kong. Danske Vidensk. Selsk. Skrift. **9**, II, 2, 235—326 (1930).

BORCH, C. VON DER: The distribution and preliminary geochemistry of modern carbonate sediments of the Coorong area, South Australia. Geochim. Cosmochim. Acta **29**, 781—799 (1965).

BORNMÜLLER, H.: Chemisch-petrographische Studien über Geoden. Jahresber. Niedersächs. Geol. Verein **16**, 28—56 (1923).

BOURGEOIS, L., TRAUBE, H.: Sur la reproduction de la dolomie. Bull. Soc. Franç. Minéral. **15**, 13—15 (1892).

BRADLEY, W. F., BURST, J. F., GRAF, D. L.: Crystal chemistry and differential thermal effects of dolomite. Am. Mineralogist **38**, 207—217 (1953).

BRADLEY, W. F., GRAF, D. L., ROTH, R. S.: The vaterite type ABO_3 rare-earth borates. Acta Cryst. **20**, 283—287 (1966).

BRADLEY, W. H., EUGSTER, H. P.: Geochemistry and paleolimnology of the trona deposits and associated minerals of the Green River formation of Wyoming. U.S. Geol. Survey Prof. Paper 496 B (1969).

BRAGG, W. H.: X-rays and crystal structure: Phil. Trans. Roy. Soc. London, Ser. A **215**, 253—274 (1915).

BRAGG, W. H., BRAGG, W. L.: X-rays and crystal structure, 2nd Ed. (1916), p. 114. London: Bell 1915.

BRAGG, W. L.: The analysis of crystals by the X-ray spectrometer. Proc. Roy. Soc. London, Ser. A **89**, 468—489 (1914).

BRAGG, W. L.: The structure of aragonite. Proc. Roy. Soc. London, Ser. A **105**, 16—39 (1924).

BRAGG, W. L.: The refractive indices of calcite and aragonite. Proc. Roy. Soc. London, Ser. A **105**, 370—386 (1924a).

BRAGG, W. L., CLARINGBULL, G. F.: Crystal structures of minerals. London: Bell 1965.

BROOKS, R., CLARK, L. M., THURSTON, E. F.: Calcium carbonate and its hydrates. Phil. Trans. Roy. Soc. London, Ser. A 243, 145—167 (1950).

BROWN, C. J., PEISER, H. S., TURNER-JONES, A.: The crystal structure of sodium sesquicarbonate. Acta Cryst. 2, 167—174 (1949).

BUBB, J. N., PERRY, D.: Porosity in some synthetic dolomites. J. Sediment. Petrol. 38, 247—249 (1968).

BUERGER, M. J.: X-ray crystallography, pp. 68—71. New York: Wiley 1942.

BUNN, C. W.: Chemical crystallography, 2nd Ed. p. 311 (1961), 1st Ed. p. 285. Oxford: Clarendon 1945.

BURLEY, G.: Polymorphism of silver iodide. Am. Mineralogist 48, 1266—1276 (1963).

BURNS, J. F., BREDIG, M. A.: Transformation of calcite to aragonite by grinding. J. Chem. Phys. 25, 1281 (1956).

BURY, C. R., REDD, R.: The system sodium carbonate–calcium carbonate–water. J. Chem. Soc. (Lond.) 1933, 1160—1162.

CARLSTRÖM, D.: A crystallographic study of vertebrate otoliths. Biol. Bull. 125, 441—463 (1963).

CHANG, L. L. Y.: Synthesis of $MBa(CO_3)_2$ compounds. Am. Mineralogist 49, 1142—1143 (1964).

CHATELIER, H. LE: Sur la chaleur de formation de l'arragonite. Compt. Rend. 116, 390—392 (1893).

CHAUDRON, G.: Contribution à l'étude des réactions dans l'état solide: Cinétique de la transformation aragonite-calcite. Proc. Intern. Symp. Reactivity Solids, Gothenburg, pp. 9—20, 1952.

CHAVE, K. E.: Aspects of the biogeochemistry of magnesium. J. Geol. 62, 266—283, 587—599 (1954).

CHAVE, K. E., DEFFEYES, K. S., WEYL, P. K., GARRELS, R. M., THOMPSON, M. E.: Observations on the solubility of skeletal carbonates in aqueous solutions. Science 137, 33—34 (1962).

CHAVE, K. E., SCHMALZ, R. F.: Carbonate-seawater interactions. Geochim. Cosmochim. Acta 30, 1037—1048 (1966).

CHESSIN, H., HAMILTON, W. C., POST, BEN: Position and thermal parameters of oxygen atoms in calcite. Acta Cryst. 18, 689—693 (1965).

CHILINGAR, G. V., BISSELL, H. J., FAIRBRIDGE, R. W. (Eds.): Carbonate rocks. Amsterdam—London—New York: Elsevier 1967.

CHRIST, C. L., HOSTETLER, P. B.: Studies in the system $MgO-SiO_2-CO_2-H_2O$ II. The activity-product constant of magnesite: Am. J. Sci. 268, 439—453 (1970).

CLARKE, F. W.: The data of geochemistry. U.S. Geol. Survey Bull. 770 (1924).

CLARKE, F. W., WHEELER, W. C.: The inorganic constituents of marine invertebrates. U.S. Geol. Survey Prof. Paper 124, 2nd ed. 1922.

CLOUD, P. E.: Environment of calcium carbonate deposition west of Andros Island, Bahamas: U.S. Geol. Survey Prof. Paper 350 (1962).

COLE, W. F., KROONE, B.: Carbonate minerals in hydrated portland cement. Nature 184, B. A., 57 (1959).

COLEMAN, R. G., LEE, D. E.: Metamorphic aragonite in the glaucophane schists of Cazadero, California. Am. J. Sci. 260, 577—595 (1962).

CORAZZA, E., SABELLI, C.: The crystal structure of pirssonite $CaNa_2(CO_3)_2 \cdot 2H_2O$. Acta Cryst. 23, 763—766 (1967).

CORNU, F.: Über die Bildungsbedingungen von Aragonit- und Kalksinter in den alten Grubenbauen der obersteirischen Erzbergwerke. Oesterr. Z. Berg. Hüttenw. 45, 596—598 (1907).

208 References

CORRENS, C. W.: Introduction to mineralogy. Berlin-Heidelberg-New York: Springer 1969.

CRAWFORD, W. A., FYEE, W. S.: Calcite-aragonite equilibrium at 100° C. Science **144**, 1569—1570 (1964).

CREDNER, H.: Ueber gewisse Ursachen der Krystallverschiedenheiten des kohlensauren Kalkes. J. prakt. Chem. **110** (2 N.F.) 292—319 (1870).

CURL, R. L.: The aragonite-calcite problem. Bull. Natl. Speleol. Soc. **24**, 57—73 (1962).

DACHILLE, F., ROY, R.: High pressure phase transformation in laboratory mechanical mixers and mortars. Nature **186**, 34, 71—72 (1960).

DAL NEGRO, A., UNGARETTI, L.: Refinement of the crystal structure of aragonite. Am. Mineralogist **56**, 768—772 (1971).

DANA, E. S.: The system of mineralogy, 6th Ed., p. 267. New York: Wiley 1892.

DAVIS, B. L., ADAMS, L. H.: Kinetics of the calcite-aragonite transformation. J. Geophys. Res. **70**, 433—441 (1965).

DEFFEYES, K. S., LUCIA, F. J., WEYL, P. K.: Dolomitization of Recent and Plio-pleistocene sediments by marine evaporite waters on Bonaire, Netherlands Antilles. Soc. Econ. Paleont. Mineral. Spec. Publ. **13**, 71—88 (1965).

DELDEN, A. VAN: Beitrag zur Kenntnis der Sulfatreduktion durch Bakterien. Centralbl. Bakt. (Abt. 2), **11**, 81—94; 113—119 (1903).

DEUSER, W. G.: Extreme $^{13}C/^{12}C$ variation in Quarternary dolomites from the continental shelf. Earth Planet. Sci. Letters **8**, 118—124 (1970).

DODD, J. R.: Magnesium and strontium in calcareous skeletons. J. Paleont. **41**, 1313—1329 (1967).

DOUGLAS, H. W., WALKER, R. A.: The electrokinetic behaviour of Iceland spar against aqueous electrolyte solutions. Trans. Faraday Soc. **46**, 559—568 (1950).

EARDLEY, A. J.: Sediments of the Great Salt Lake, Utah. Bull. Am. Ass. Petroleum Geol. **22**, 1359—1387 (1938).

ENGELHARDT, W. V.: Der Porenraum der Sedimente. Berlin-Göttingen-Heidelberg: Springer 1960.

ENGELHARDT, W. V.: Interstitial solutions and diagenesis in sediments. In: Developments in Sedimentology 8, Diagenesis in Sediments, pp. 503—521. Amsterdam—London—New York: Elsevier 1967.

ENGELHARDT, W. V.: Die Bildung von Sedimenten und Sedimentgesteinen. Stuttgart: Schweizerbart 1973.

ERENBURG, B. G.: Artificial mixed carbonates in the $CaCO_3$-$MgCO_3$ series. Zhurnal Strukt. Khim. **2**, 178—182 (1961).

EWALD, P. P., HERRMANN, C.: Structurbericht 1913—1928, pp. 295—297. Leipzig 1931.

FAHEY, J. J. (MROSE, M. E.): Saline minerals from the Green River formation. U.S. Geol. Survey Prof. Paper 405 (1962).

FAHEY, J. J., YORKS, K. P.: Wegscheiderite ($Na_2CO_2 \cdot 3NaHCO_3$), a new saline mineral from the Green River formation, Wyoming. Am. Mineralogist **48**, 400—403 (1963).

FAIRBRIDGE, R. W.: The dolomite question. Soc. Econ. Paleont. Mineralog. Spec. Publ. **5**, 125—178 (1957).

FAUST, G. T.: Huntite, $Mg_3Ca(CO_3)_4$, a new mineral. Am. Mineralogist **38**, 4—24 (1953).

FAUST, G. T.: Thermal analysis studies on carbonates I, aragonite and calcite. Am. Mineralogist **35**, 207—224 (1950).

FISCHBECK, R., MÜLLER, G.: Monohydrocalcite, hydromagnesite, nesquehonite, dolomite, aragonite, and calcite in speleothems of the Fränkische Schweiz, Western Germany. Contrib. Mineral. Petrol. 33, 87—92 (1971).

FISCHER, A. G., GARRISON, R. E.: Carbonate lithification on the sea floor. J. Geol. 75, 488—496 (1967).

FLÖRKE, W., FLÖRKE, O. W.: Vateritbildung aus Gips in Sodalösung. Neues Jahrb. Mineral. Monatsh. 1961, 179—181.

FLÜGEL, H. W., WEDEPOHL, K. H.: Die Verteilung des Strontiums in oberjurassischen Karbongesteinen der Nördlichen Kalkalpen. Contrib. Mineral. Petrol. 14, 229—249 (1967).

FORCHHAMMER, G.: Beiträge zur Bildungsgeschichte des Dolomits. Neues Jahrb. Mineral. 1852, 854—858.

FRIEDMAN, G. M: Early diagenesis and lithification in carbonate sediments. J. Sediment. Petrol. 34, 777—813 (1964).

FRIEDMAN, G. M.: The fabric of carbonate cement and its dependence on the salinity of water. In: MÜLLER, FRIEDMAN (Eds.): Recent developments in carbonate sedimentology in Central Europe, pp. 11—20. Berlin-Heidelberg-New York: Springer 1968.

FROESE, E.: A note on strontium magnesium carbonate. Can. Mineralogist 9, 65—70 (1967).

FRONDEL, C., BAUER, L. H.: Kutnahorite: a manganese dolomite, CaMn(CO₃)₂. Am. Mineralogist 40, 748—760 (1955).

FRUEH, JR. A. J., GOLIGHTLY, J. P.: The crystal structure of dawsonite, NaAl(CO₃)(OH)₂. Can. Mineralogist 9, 51—56 (1967).

FÜCHTBAUER, H., GOLDSCHMIDT, H.: Aragonitische Lumachellen im bituminösen Wealden des Emslandes. Beitr. Mineral. Petrogr. 10, 184—197 (1964).

FÜCHTBAUER, H., GOLDSCHMIDT, H.: Beziehungen zwischen Calciumgehalt und Bildungsbedingungen der Dolomite. Geol. Rundschau 55, 29—40 (1965).

FÜCHTBAUER, H., MÜLLER, G.: Sedimente und Sedimentgesteine. Stuttgart: Schweizerbart 1970.

FYFE, W. S., BISCHOFF, J. L.: The calcite-aragonite problem. Soc. Econ. Paleont. Mineral., Spec. Public. 13, 3—13 (1965).

GARRELS, R. M., CHRIST, C. L.: Solutions, minerals and equilibria. New York: Harper & Row 1965.

GARRELS, R. M., THOMPSON, M. E., SIEVER, R.: Stability of some carbonates at 25° C and one atmosphere total pressure. Am. J. Sci. 258, 402—418 (1960).

GARRELS, R. M., THOMPSON, M. E., SIEVER, R.: Control of carbonate solubility by carbonate complexes. Am. J. Sci. 259, 24—45 (1961).

GARRELS, R. M., THOMPSON, M. E.: A chemical model for sea water at 25° C and one atomosphere total pressure. Am. J. Sci. 260, 57—66 (1962).

GAVISH, E., FRIEDMAN, G. M.: Progressive diagenesis in Quaternary to late Tertiary carbonate sediments: sequence and time scale. J. Sediment. Petrol. 39, 980—1006 (1969).

GEE, H., REVELLE, R.: Calcium equilibrium in sea water. Bull. Scripps Inst. Oceanogr. 3, 180—190 (1932).

GEVIRTZ, J. L., FRIEDMAN, G. M.: Deep-sea carbonate sediments of the Red Sea and their implication on marine carbonate lithification. J. Sediment. Petrol. 36, 143—151 (1966).

GINSBURG, R. N.: Beachrock in South Florida. J. Sediment. Petrol. 23, 85—92 (1953).

GINSBURG, R. N.: Early diagenesis and lithification of shallow-water carbonate sediments in South Florida. Soc. Econ. Paleont. Mineral. Spec. Publ. **5**, 80—99 (1957).

GLOVER, E. D., SIPPEL, R. F.: Experimental pseudomorphs: replacement of calcite by fluorite. Am. Mineralogist **47**, 1156—1165 (1962).

GLOVER, E. D., SIPPEL, R. F.: Synthesis of magnesium calcite. Geochim. Cosmochim. Acta **31**, 603—613 (1967).

GOLDSCHMIDT, V. M., HAUPTMANN, H.: Isomorphie von Boraten und Karbonaten. Nachr. Ges. Wissensch. Göttingen, Math. phys. Kl. **1931—1932**, 53—72.

GOLDSMITH, J. R., GRAF, D. L.: The system CaO-MnO-CO$_3$; solid solution and decomposition relations. Geochim. Cosmochim. Acta **11**, 310—334 (1957).

GOLDSMITH, J. R., GRAF, D. L.: Structural and compositional variations in some natural dolomites. J. Geol. **66**, 678—692 (1958).

GOLDSMITH, J. R., GRAF, D. L.: Relation between lattice constants and composition of the Ca-Mg carbonates. Am. Mineralogist **43**, 84—101 (1958 a).

GOLDSMITH, J. R., GRAF, D. L., HEARD, H. C.: Lattice constants of the calcium-magnesium carbonates. Am. Mineralogist **46**, 453—457 (1961).

GOLDSMITH, J. R., GRAF, D. L., JOENSUU, O. I.: The occurrence of magnesian calcites in nature. Geochim. Cosmochim. Acta **7**, 212—230 (1955).

GOLDSMITH, J. R., GRAF, D. L., WITTERS, J., NORTHROP, D. A.: Studies in the system CaCO$_3$-MgCO$_3$-FeCO$_3$. J. Geol., **70**, 659—688 (1962).

GOLDSMITH, J. R., HEARD, H. C.: Subsolidus phase relations in the system CaCO$_3$-MgCO$_3$. J. Geol. **69**, 45—74 (1961).

GOLDSMITH, J. R., NEWTON, R. C.: P–T–X relations in the system CaCO$_3$-MgCO$_3$ at high temperatures and pressures. Am. J. Sci. **267-A**, 160—190 (1969).

GORDON, L., SALUTSKY, M. L., WILLARD, H. H.: Precipitation from homogeneous solution. New York: Wiley 1960.

GOSSNER, B.: Aragonitgruppe. In: Handb. Mineral. von HINTZE, C., Bd. **1**, Abt. 3, 1. Teil, S. 2958—3111. Berlin–Leipzig: de Gruyter 1930.

GOTO, M.: Some mineralo-chemical problems concerning calcite and aragonite, with special reference to the genesis of aragonite. J. Fac. Sci. Hokkaido Univ., Ser. IV Geol. Mineral. **10**, 571—640 (1961).

GRAF, D. L.: Geochemistry of carbonate sediments and sedimentary carbonate rocks. Illinois State Geol. Survey Circ. 297; 298; 301; 308; 309 (1960).

GRAF, D. L.: Crystallographic tables for the rhombohedral carbonates. Am. Mineralogist **46**, 1283—1316 (1961).

GRAF, D. L.: Crystallographic tables for the rhombohedral carbonates: a correction. Am. Mineralogist **54**, 325 (1969).

GRAF, D. L., BLYTH, C. R., STEMMLER, R. S.: One-dimensional disorder in carbonates. Illinois State Geol. Survey Circ. 408 (1967).

GRAF, D. L., BRADLEY, W. F.: The crystal structure of huntite, Mg$_3$Ca(CO$_3$)$_4$. Acta Cryst. **15**, 238—242 (1962).

GRAF, D. L., EARDLEY, A. J., SHIMP, N. F.: A preliminary report on magnesium carbonate formation in Glacial Lake Bonneville. J. Geol. **69**, 219—223 (1961).

GRAF, D. L., GOLDSMITH, J. R.: Some hydrothermal syntheses of dolomite and protodolomite. J. Geol. **64**, 173—186 (1956).

GRAF, D. L., LAMAR, J. E.: Properties of calcium and magnesium carbonates and their bearing on some uses of carbonates rocks. Econ. Geol. 50th annivers. vol. 639—713 (1955).

GREENWALD, I.: The dissociation of calcium and magnesium carbonates and bicarbonates. J. Biol. Chem. **141**, 789 (1941).

GROSS, R., MÖLLER, H.: Über das Kristallwachstum in röhrenförmigen Hohlräumen. Z. Physik **19**, 374—375 (1923).

HALLA, F., RITTER, F.: Eine Methode zur Bestimmung der freien Energie bei Reaktionen des Typus A(s) + B(s) — AB(s) und ihre Anwendung auf das Dolomitproblem. Z. phys. Chem. **175A**, 63—82 (1935).

HALLA, F., TASSEL, R. VAN: Löslichkeitsanomalien bei Magnesit. Radex-Rundschau **1964**, 42—44.

HALLAM, A., O'HARA, M. J.: Aragonitic fossils in the lower Carboniferous of Scotland. Nature **195**, 273—274 (1962).

HARDIE, L. A.: The gypsum-anhydrite equilibrium at one atmosphere pressure. Am. Mineralogist **52**, 171—200 (1967).

HARPER, J. P.: Crystal structure of sodium carbonate monohydrate, $NaCO_3 \cdot H_2O$. Z. Krist. **95**, 266—273 (1936).

HARVEY, K. B., PORTER, G. B.: Introduction to physical inorganic chemistry. Reading, Mass.: Addison-Wesley 1963.

HEIDE, F.: Über den Vaterit. Centralbl. Mineral. **1924**, 641—651.

HELING, D.: Microporosity of carbonate rocks. In: MÜLLER, FRIEDMANN (Eds.): Recent developments in carbonate sedimentology in Central Europe, pp. 98—105. Berlin-Heidelberg-New York: Springer 1968.

HINTZE, C.: Handbuch der Mineralogie. Nitrate, Jodate, Karbonate etc., Bd. 1, Abt. 3, 1. Teil. Berlin—Leipzig: de Gruyter 1930.

HOEFS, J.: Kohlenstoff- und Sauerstoff-Isotopenuntersuchungen an Karbonatkonkretionen und umgebendem Gestein. Contrib. Mineral. Petrol. **27**, 66—79 (1970).

HORN, G.: Löslichkeitskonstanten und Freie Bildungsenthalpien von Magnesit, Brucit und Dolomit. Radex-Rundschau **1969**, 439—459.

HOROWITZ, A. S., POTTER, P. E.: Introductory petrography of fossils. New York—Heidelberg Berlin: Springer 1971.

HOWIE, R. A., BROADHURST, F. M.: X-ray data for dolomite and ankerite. Am. Mineralogist **43**, 1210—1214 (1958).

HSU, K. J.: Chemistry of dolomite formation. In: CHILINGAR, G. V. et al. (Eds.): Carbonate rocks, B, pp. 169—191. Amsterdam: Elsevier 1967.

ILLING, L. V.: Bahaman calcareous sands. Bull. Am. Ass. Petrol. Geol. **38**, 1—95 (1954).

ILLING, L. V., WELLS, A. J., TAYLOR, J. C. M.: Penecontemporary dolomite in the Persian Gulf. Soc. Econ. Paleont. Mineral. Spec. Publ. **13**, 89—111 (1965).

INKINEN, O., LAHTI, L.: X-ray crystal analysis of calcite: Ann. Acad. Sci. Fennicae. Ser. A., VI Phys. No. 141—142 (1964).

International Tables for X-ray Crystallography: vol. I (1952); vol. II (1959); Birmingham (England): Kynoch.

IRION, G., MÜLLER, G.: Mineralogy, petrology and chemical composition of some calcareous tufa from the Schwäbische Alb, Germany. In: MÜLLER, FRIEDMAN (Eds.): Recent developments in carbonate sedimentology in Central Europe, pp. 157—171. Berlin-Heidelberg-New York: Springer 1968.

JAGODZINSKI, H.: Kristallstruktur und Fehlordnung des Artinits, $Mg_2(CO_3(OH)_2) \cdot 3H_2O$. Tschermaks Mineral. Petrogr. Mitt. **10**, 297—330 (1965).

JAMIESON, J. C.: Phase equilibrium in the system calcite-aragonite: J. Chem. Phys. **21**, 1385—1390 (1953).

JAMIESON, J. C.: Introductory studies of high-pressure polymorphism to 24,000 bars by X-ray diffraction with some comment on calcite II. J. Geol. **65**, 334—343 (1957).

JAMIESON, J. C., GOLDSMITH, J. R.: Some reactions produced in carbonates by grinding. Am. Mineralogist **45**, 818—827 (1960).

JANSEN, J. F., KITANO, Y.: The resistance of Recent marine carbonate sediments to solution. J. Oceanogr. Soc. Japan **18**, (42)—(53), 208—219 (1963).

JANTSCH, G., ZEMEK, F.: Die Darstellung von synthetischem kristallinen Magnesit. Radex Rundschau **3**, 110—111 (1949).

JIRGENSONS, B., STRAUMANIS, M. E.: A short textbook of colloid chemistry, 2nd Ed. Oxford: Pergamon 1962.

JOHNSTON, J., MERWIN, H. E., WILLIAMSON, E. D.: The several forms of calcium carbonate. Am. J. Sci. Ser. 4, **41**, 473—512 (1916).

JOOS, G.: Theoretical physics. London: Blackie 1934.

KAMHI, S. R.: On the structure of vaterite, $CaCO_3$. Acta Cryst. **16**, 770—772 (1963).

KAPUSTIN, YU. L.: Norsethite, the first locality in the USSR (in Russian). Dokl. Akad. Nauk. SSSR **161**, 922—924 (1965).

KATZ, A.: Calcian dolomites and dedolomitization. Nature **217**, 439—440 (1968).

KEESTER, K. L., JOHNSON, JR., G. G., VAND, V.: New data on tychite. Am. Mineralogist **54**, 302—305 (1969).

KEISER (KEYZER), N.: Materialien zur Geschichte, Morphologie und Hydrologie des Sees Issyk-Kul. Acta Univ. Asiae Mediae, Ser. 12a, Geogr. **1**, 39—43 (1928).

KEYSER, W. L. DE, DEGUELDRE, L.: Contribution à l'étude de la formation de la calcite, aragonite et vatérite. Bull. Soc. Chim. Belg. **59**, 40—71 (1950).

KINSMAN, D. J. J.: Huntite from a carbonate evaporite environment. Am. Mineralogist **52**, 1332—1340 (1967).

KINSMAN, D. J. J.: Interpretation of Sr^{2+} concentrations in carbonate minerals and rocks. J. Sediment. Petrol. **39**, 486—508 (1969).

KITANO, Y.: A study of the polymorphic formation of calcium carbonate in thermal springs with an emphasis on the effect of temperature. Bull. Chem. Soc. Japan **35**, 1980—1985 (1962a).

KITANO, Y.: The behavior of various inorganic ions in the separation of calcium carbonate from a bicarbonate solution. Bull. Chem. Soc. Japan **35**, 1973—1980 (1962b).

KITANO, Y., HOOD, D. W.: Calcium carbonate crystal forms formed from sea water by inorganic processes. J. Oceanogr. Soc. Japan **18**, (35)—(39), 141—145 (1962).

KITANO, Y., HOOD, D. W.: The influence of organic material on the polymorphic crystallization of calcium carbonate. Geochim. Cosmochim. Acta **29**, 29—41 (1965).

KITANO, Y., KANAMORI, N.: Synthesis of magnesian calcite at low temperatures and pressures. Geochem. J. (Japan) **1**, 1—10 (1966).

KITANO, Y., KAWASAKI, N.: Behavior of strontium ion in the process of calcium carbonate separation from bicarbonate solution. J. Earth Sci. Nagoya Univ. **6**, 63—74 (1958).

KNATZ, H.: Mineralogisch-petrographische Untersuchungen an künstlich gebildeten Kalkooiden. Glückauf-Forschungsh. **26**, 169—175 (1965).

KNATZ, H.: Zur Bildung „künstlicher Ooide" in Kraftwerken. Leitz-Mitt. Wissensch. Techn. **3**, 176—178 (1966).

KORTÜM, G., HAUG, P.: Thermodynamische Untersuchungen am System 2,4-Lutidin-Wasser mit geschlossener Mischungslücke. Z. Elektrochem.-Ber. Bunsenges. Phys. Chem. **60**, 355—362 (1956).

KROTOV, B. P.: Dolomite, ihre Bildung, Existenzbedingungen in der Erdkruste und Umwandlungen im Zusammenhang mit dem Studium des oberen Teils der Kazan-Stufe in der Umgebung von Kazan. (In Russian with an extended summary in German). Verh. Naturforsch. Ges. Univ. Kazan, **50** Lief. 6, 1—110 (1925), abstract. Neu. Jahrb. Mineral. Geol. II B, 292 (1926).

KÜBLER, B.: Calcites magnésiennes d'eau douce dans le Tertiaire supérieur du Jura neuchâtelois (Suisse). Eclogae Geol. Helv. **51**, 676—685 (1958).

KÜBLER, B.: Etude pétrographique de l'Oehningien (Tortonien) du Locle (Suisse occidentale). Beitr. Mineral. Petrogr. **8**, 267—314 (1962).

LAND, L. S., GOREAU, T. F.: Submarine lithification of Jamaican reefs. J. Sediment. Petrol. **40**, 457—462 (1970).

LANDER, J. J.: Polymorphism and anion rotational disorder in the alkaline earth carbonates. J. Chem. Phys. **17**, 892—901 (1949).

LANGMUIR, D.: Stability of carbonates in the system $MgO-CO_2-H_2O$. J. Geol. **73**, 730—754 (1965).

LEITMEIER, H.: Zur Kenntnis der Carbonate I Die Dimorphie des kohlensauren Kalkes. Neues Jahrb. Mineral. **1910**, 49—77.

LEITMEIER, H.: Zur Kenntnis der Carbonate II. Neues Jahrb. Mineral. Beil.-Bd. **40**, 655—700 (1916).

LENNARD-JONES, J. E., DENT, B. M.: Some theoretical determinations of the structure of carbonate crystals. Proc. Roy. Soc. London, Ser. A **113**, 673—696 (1927).

LERMAN, A.: Paleoecological problems of Mg and Sr in biogenic calcites in the light of recent thermodynamic data. Geochim. Cosmochim. Acta **29**, 977—1002 (1965).

LEVIN, E. M., ROTH, R. S., MARTIN, J. B.: Polymorphism of ABO_3 type rare-earth borates. Am. Mineralogist **46**, 1030—1055 (1961).

LINCK, G.: Die Bildung der Oolithe und Rogensteine. Neues Jahrb. Mineral. Beil.-Bd. **16**, 495—513 (1903).

LIPPMANN, F.: Ton, Geoden und Minerale des Barrême von Hoheneggelsen. Geol. Rundschau **43**, 475—503 (1955).

LIPPMANN, F.: Darstellung und kristallographische Eigenschaften von $CaCO_3 \cdot H_2O$. Naturwissenschaften **46**, 553—554 (1959).

LIPPMANN, F.: Versuche zur Aufklärung der Bildungsbedingungen von Calcit und Aragonit. Fortschr. Mineral. **38**, 156—161 (1960).

LIPPMANN, F.: Benstonite, $Ca_7 Ba_6 (CO_3)_2$, a new mineral from the barite deposit in Hot Spring County, Arkansas. Am. Mineralogist **47**, 585—598 (1962).

LIPPMANN, F.: $PbMg(CO_3)_2$, ein neues rhomboedrisches Doppelkarbonat. Naturwissenschaften **53**, 701 (1966).

LIPPMANN, F.: Die Synthese des Norsethit bei 20° C und 1 at. Ein Modell zur Dolomitisierung. Neues Jahrb. Mineral. Monatsh. **1967**, 23—29.

LIPPMANN, F.: Die Kristallstruktur des Norsethit, $BaMg(CO_3)_2$, mit einem Strukturvorschlag für $PbMg(CO_3)_2$. Tschermaks Mineral. Petrogr. Mitt. **12**, 299—318 (1968a).

LIPPMANN, F.: Syntheses of $BaMg(CO_3)_2$ (Norsethite) at 20° C and the formation of dolomite in sediments. In: MÜLLER, FRIEDMAN (Eds.): Recent developments in carbonate sedimentology in Central Europe, pp. 33—37. Berlin-Heidelberg-New York: Springer 1968b.

LIPPMANN, F., JOHNS, W. D.: Regular interstratification in rhombohedral carbonates and layer silicates. Neues Jahrb. Mineral. Monatsh. **1969**, 212—221.

LIPPMANN, F., SCHLENKER, B.: Mineralogische Untersuchungen am Oberen Muschelkalk von Haigerloch (Hohenzollern). Neues Jahrb. Mineral. Abhandl. **113**, 68—90 (1970).

LIVINGSTONE, D. A.: Chemical composition of rivers and lakes. U.S. Geol. Survey Prof. Paper 440 G (1963).

LOWENSTAM, H. A.: Factors affecting the aragonite:calcite ratios in carbonate-secreting marine organisms. J. Geol. **62**, 284—322 (1954).

LOWENSTAM, H. A.: Aragonite needles secreted by algae and some sedimentary implications. J. Sediment. Petrol. **25**, 270—272 (1955).

MACDONALD, G. J. F.: Experimental determination of calcite-aragonite equilibrium relations at elevated temperature and pressures. Am. Mineralogist **41**, 744—756 (1956).

MACINTYRE, I. G., MOUNTJOY, E. W., D'ANGLEJAN, B. F.: An occurrence of submarine cementation of carbonate sediments off the west coast of Barbados, W. I.. J. Sediment. Petrol. **38**, 660—664 (1968).

MACKENZIE, F. T., GARRELS, R. M.: Silica-bicarbonate balance in the ocean and early diagenesis. J. Sediment. Petrol. **36**, 1075—1084 (1966).

MARC, R., ŠIMEK, A.: Über die thermische Dissoziation des Magnesiumkarbonats. Z. Anorg. Chem. **82**, 17—49 (1913).

MATTAVELLI, L.: Osservazioni petrografiche sulla sostituzione della dolomite con la calcite (dedolomitizzazione) in alcune facies carbonate italiane. Atti Soc. Ital. Sci. Nat. Mus. Civ. Stor. Nat. Milano **105**, 294—316 (1966).

MATTHEWS, R. K.: Carbonate diagenesis: equilibration of sedimentary mineralogy to the subaerial environment; coral cap of Barbados, West Indies. J. Sediment. Petrol. **38**, 1110—1119 (1968).

MAYER, F. K., WEINECK, E.: Die Verbreitung des Kalziumkarbonat im Tierreich unter besonderer Berücksichtigung der Wirbellosen. Jena. Z. Naturwiss. **66**, (59), 199—222 (1932).

MCCONNELL, J. D. C.: Vaterite from Ballycraigy, Larne, Northern Ireland. Mineral. Mag. **32**, 535—544 (1960).

MCKEE, B.: Aragonite in the Franciscan rocks of the Pacheco Pass area, California. Am. Mineralogist **47**, 379—387 (1962).

MEDLIN, W. L.: The preparation of synthetic dolomite. Am. Mineralogist **44**, 979—986 (1959).

MEIGEN, W.: Beiträge zur Kenntnis des kohlensauren Kalkes. Ber. Naturforsch. Ges. Freiburg i. Br. **13**, 40—94 (1903).

MEIGEN, W.: Beiträge zur Kenntnis des kohlensauren Kalkes II, III. Ber. Naturforsch. Ges. Freiburg i. Br. **15**, 38—54, 55—74 (1907).

MEIGEN, W.: Über kohlensauren Kalk. Verhandl. Ges. Deut. Naturforsch. Ärzte **82**, II, 1, 120—124 (1911).

MENCHETTI, S.: Dati cristallografici-strutturali sulla gaylussite. Rend. Soc. Mineral. Ital. **24**, 277—281 (1968).

MEYER, H. J.: Über Vaterit und seine Struktur. Fortschr. Mineral. **38**, 186—187 (1960).

MEYER, H. J.: Struktur und Fehlordnung des Vaterits. Z. Krist. **128**, 183—212 (1969).

MILLIMAN, J. D., ROSS, D. A., TEH-LUNG KU: Precipitation and lithification of deep-sea carbonate in the Red Sea. J. Sediment. Petrol. **39**, 724—736 (1969).

MILLS, A. D.: Crystallographic data for new rare-earth borate compounds, $RX_3(BO_3)_4$. Inorgan. Chem. **1**, 960—961 (1962).

MILTON, C., EUGSTER, H. P.: Mineral assemblages of the Green River formation. In: ABELSON, P. H. (Ed.): Researches in Geochemistry, pp. 118—150. New York: Wiley 1959.

MILTON, C., FAHEY, J. J.: Classification and association of the carbonate minerals of the Green River formation. Am. J. Sci. **258-A**, 242—246 (1960).

MONAGHAN, P. H., LYTLE, M. L.: The origin of calcareous ooliths. J. Sediment. Petrol. **26**, 111—118 (1956).

MORANDI, N.: La dissociazione termica dell' idromagnesite e della nesquehonite. Mineral. Petrogr. Acta **15**, 93—108 (1969).

MORELLI, G. L.: Determinazione della composizione delle fasi trigonali nel sistema $MgCO_3$-$FeCO_3$-$CaCO_3$ mediante diffrazione dei raggi X. Rend. Soc. Mineral. Ital. **23**, 315—332 (1967).

MROSE, M. E., CHAO, E. C. T., FAHEY, J. J., MILTON, C.: Norsethite, $BaMg(CO_3)_2$, a new mineral from the Green River Formation, Wyoming. Am. Mineralogist **46**, 420—429 (1961).

MÜLLER, G., FRIEDMAN, G. M. (Eds.): Recent developments in carbonate petrology in Central Europe. Berlin-Heidelberg-New York: Springer 1968.

MÜLLER, G., TIETZ, G.: Recent dolomitization of Quaternary biocalcarenites from Fuerteventura (Canary Islands). Contrib. Mineral. Petrol. **13**, 89—96 (1966).

MURATA, K. J., ERD, R. C.: Composition of sediment from the experimental Mohole project (Guadalupe site). J. Sediment. Petrol. **34**, 633—655 (1964).

MURATA, K. J., FRIEDMAN, I., MADSEN, B. M.: Isotopic composition of diagenetic marine carbonates in marine Miocene formations of California and Oregon. U.S.Geol. Survey Prof. Paper 614 B (1969), see also: U.S.Geol. Survey Prof. Paper 724 C (1972).

MURRAY, J. W.: The deposition of calcite and aragonite in caves. J. Geol. **62**, 481—492 (1954).

NASHAR, B.: Barringtonite — A new hydrous magnesium carbonate from Barrington Tops, New South Wales, Australia. Mineral. Mag. **34**, 370—372 (1965).

NEEV, D., EMERY, K. O.: The Dead Sea. Israel Geol. Survey Bull. 41 (1967).

NEWELL, N. D., PURDY, E. G., IMBRIE, J.: Bahamian oölitic sand. J. Geol. **68**, 481—497 (1960).

NOYES, R. M.: Thermodynamics of ion hydration as a measure of effective dielectric properties of water. J. Am. Chem. Soc. **84**, 513—522 (1962).

OLBY, J. K.: The basic lead carbonates. J. Inorg. Nucl. Chem. **28**, 2507—2512 (1966).

OPPENHEIMER, C. H.: Note on the formation of spherical aragonitic bodies in the presence of bacteria from the Bahama Bank. Geochim. Cosmochim. Acta **23**, 295—296 (1961).

OLSHAUSEN, S. v.: Strukturuntersuchungen nach der Debye-Scherrer-Methode. Z. Krist. **61**, 463—514 (1925).

OVERBEEK, J. T. G.: In: KRUYT, H. R. (Ed.): Colloid Science Vol. 1. Amsterdam: Elsevier 1952.

PALACHE, C., BERMAN, H., FRONDEL, C.: Dana's system of mineralogy, Vol. II. New York: Wiley 1951.

PHEMISTER, D. B., ARONSOHN, H. G., PEPINSKY, R.: Variations in the cholesterol bile pigment and the calcium salt contents of gallstones formed in gall-bladder and in bile ducts with the degree of associated obstruction. Ann. Surg. **109**, 161—186 (1939).

PEDERSEN, B.: The structure of the $(HCO_3)_2^{2-}$ ion in potassium bicarbonate. Acta Cryst., **B 24**, 478—480 (1968).

PIERCE, J. W., MELSON, W. G.: Dolomite from the continental slope off southern California. J. Sediment. Petrol. **37**, 963—966 (1967).

PILKEY, O. H., BLACKWELDER, B. W.: Mineralogy of the sand size carbonate fraction of some recent marine terrigenous carbonate sediments. J. Sediment. Petrol. **38**, 799—810 (1968).

PINGITORE, N. E. JR.: Diagenesis and porosity modification in Acropora palmata, Pleistocene of Barbados, West Indies. J. Sediment. Petrol. **40**, 712—721 (1970).

POBEGUIN, TH.: Contribution à l'étude des carbonates de calcium. Précipitation du calcaire par les végétaux. Comparison avec le monde animal. Ann. Sci. Nat. Botan. 11 sér. **15**, 29—109 (1954).

PRIEN, F. L., FRONDEL, C.: Studies in urolithiasis: I. The composition of urinary calculi. J. Urol. **57**, 949—994 (1947).

PURDY, E. G.: Recent calcium carbonate facies of the Great Bahama Bank. J. Geol. **71**, 334—355, 472—497 (1963).

RAADE, G.: Dypingite, a new hydrous basic carbonate of magnesium from Norway. Amer. Mineralogist **55**, 1457—1465 (1970).

RAISTRICK, R.: The influence of foreign ions on crystal growth from solution. The stabilization of the supersaturation of calcium carbonate solution by anions possessing O–P–O–P–O chains. Disc. Faraday Soc. **5**, 234—237 (1949).

RINNE, F.: Röntgenographische Untersuchungen an einigen feinzerteilten Mineralien, Kunstprodukten und dichten Gesteinen. Z. Krist. **60**, 55—69 (1924).

ROBIE, R. A., WALDBAUM, D. R.: Thermodynamic properties of minerals and related substances at 298.15°K (25.0° C) and one atmosphere (1.013 bars) pressure and at higher temperatures. U.S.Geol. Survey Bull. 1259 (1968).

ROBINSON, R. B.: Diagenesis and porosity development in Recent and Pleistocene oolites from Southern Florida and the Bahamas. J. Sediment. Petrol. **37**, 355—364 (1967).

ROSE, G.: Ueber die Bildung des Kalkspaths and des Arragonits. Poggendorfs Ann. Phys. **42**, 353—367 (1837).

ROSE, G.: Über die heteromorphen Zustände der kohlensauren Kalkerde (Erste Abh.) Abhandl. Kön. Akad. Wiss. Berlin **1856**, 1—76.

ROTH, R. S., WARING, J. L., LEVIN, E. M.: Polymorphism of ABO_3 rare earth borate solid solutions. Rare Earth Research, Vol. II, pp. 153—163. New York: Gordon & Breach 1964.

ROTHE, P.: Dolomitization of biocalcarenites of late Tertiary age. In: MÜLLER, FRIEDMAN (Eds.): Recent developments in carbonate sedimentology in Central Europe, pp. 38—45. Berlin-Heidelberg-New York: Springer 1968.

RUMANOVA, I. M., MALITSKAYA, G. I.: Revision of the structure of astrakhanite by weighted phase projection methods. Soviet Phys.-Cryst. **4**, 481—495 (1960).

RUSNAK, G. A.: Some observations of recent oolites. J. Sediment. Petrol. **30**, 471—480 (1960).

RUSSELL, K. L., DEFFEYES, K. S., FOWLER, G. A., LLOYD, R. M.: Marine dolomite of unusual composition. Science, **155**, 189—191 (1967).

SANDS, D. E.: Introduction to crystallography, pp. 59—63. New York—Amsterdam: Benjamin 1969.

SAPOZHNIKOV, D. G., TSVETKOV, A. I.: The secretion of hydrated calcium carbonate on the bottom of Lake Issyk-Kul (in Russian): Dokl. Akad. Nauk SSSR **124**, 402—405 (1959).

SAPOZHNIKOV, D. G., VISELKINA, M. A.: Recent sediments of Lake Issyk-Kul and its bays (in Russian). Trudy Inst. Geol. Petrogr. Mineral. Geokh. No. 36, Izd. Akad. Nauk SSSR, Moscow (1960).

SASS, R. L., SCHEUERMANN, R. F.: The crystal structure of sodium bicarbonate. Acta Cryst. **15**, 77—81 (1962).

SAVELLI, C., WEDEPOHL, K. H.: Geochemische Untersuchungen an Sinterkalken (Travertinen). Contrib. Mineral. Petrol. **21**, 238—256 (1969).

SCHLANGER, S. O.: Subsurface geology of Eniwetok Atoll. U.S.Geol. Survey Prof. Paper 260-BB (1963).

SCHIEBOLD, E.: Die Verwendung der Lauediagramme zur Bestimmung der Struktur des Kalkspates. Abhandl. Math.-Phys. Kl. Sächs. Akad. Wiss. **36**, 67(70)—213 (1919).

SCHMALZ, R. F.: Brucite in carbonate secreted by the red alga Goniolithon sp. Science **149**, 993—996 (1965).

SCHOPF, T. J. M., MANHEIM, F. T.: Chemical composition of Ectoprocta (Bryozoa). J. Paleont. **41**, 1197—1225 (1967).

SCHRÖDER, F.: Vaterit, das metastabile Calciumkarbonat, als sekundäres Zementmineral. Tonind. Z. **86**, 254—260 (1962).

SEMENOV, E. I.: Hydrated carbonates of sodium and calcium: Soviet Phys.-Cryst. **9**, 88—90, Abstr. Am. Mineralogist **49**, 1151 (1964).

SHINN, E. A.: Submarine lithification of Holocene carbonate sediments in the Persian Gulf. Sedimentology **12**, 109—144 (1969).

SHINN, E. A., GINSBURG, R. N., LLOYD, R. M.: Recent supratidal dolomite from Andros Island, Bahamas. Soc. Econ. Paleont. Mineral. Spec. Publ. **13**, 112—123 (1965).

SHINN, E. A., LLOYD, R. M., GINSBURG, R. N.: Anatomy of a modern carbonate tidal-flat, Andros Island, Bahama. J. Sediment. Petrol. **39**, 1202—1228 (1969).

SIEGEL, F. R.: Factors influencing the precipitation of dolomitic carbonates. State Geol. Survey Kansas Bull. **152**, 127—158 (1961).

SIMKISS, K.: Variations in the crystalline form of calcium carbonate precipitated from artificial sea water. Nature **201**, 492—493 (1964).

SMIT, D. E., SWETT, K.: Devaluation of "dedolomitization": J. Sediment. Petrol. **39**, 379—380 (1969).

SMITH, J. W., MILTON, C.: Dawsonite in the Green River formation. Econ. Geol. **61**, 1029—1042 (1966).

SMYTHE, J. A., DUNHAM. K. C.: Ankerites and chalybdites from the northern Pennine ore field and the north-east coalfield. Mineral. Mag. **28**, 53—74 (1947).

SOGNNAES, R. F. (Ed.): Calcification in biological systems. Washington, D.C.: Am. Ass. Advan. Sci. 1960.

SORBY, H. C.: The structure and origin of limestones. Proc. Geol. Soc. London **35**, 56—95 (1879).

SPANGENBERG, K.: Die künstliche Darstellung des Dolomits. Z. Kryst. **52**, 529—567 (1913).

SPANGENBERG, K.: Wachstum und Auflösung der Kristalle: Handwörterbuch der Naturwissenschaften, Bd. 10, S. 388—390, 2. Aufl. Jena: Gustav Fischer 1935.

SPOTTS, J. H., SILVERMAN, S. R.: Organic dolomite from Point Fermin, California. Am. Mineralogist **51**, 1144—1155 (1966).

STARK, J.: Neuere Ansichten über die zwischen- und innermolekulare Bindung in Kristallen. Jahrb. Radioaktivität u. Elektr. **12**, 279—296 (1915).

STEINFINK, H., SANS, F. J.: Refinement of the crystal structure of dolomite. Am. Mineralogist **44**, 679—682 (1959).

STEHLI, F. G.: Shell mineralogy in paleozoic invertebrates. Science **123**, 1031—1032 (1956).

STEHLI, F. G., HOWER, J.: Mineralogy and early diagenesis of carbonate sediments. J. Sediment. Petrol. **31**, 358—371 (1961).

STEYN, J. G. D., WATSON, M. D.: Notes on a new occurrence of norsethite, $BaMg(CO_3)_2$. Am. Mineralogist **52**, 1770—1775 (1967).

STODDART, D. R., CANN, J. R.: Nature and origin of beachrock. J. Sediment. Petrol. **35**, 243—247 (1965).

STOJCIC, B.: Die Erdöllagerstätte von Lacq supérieur. Erdöl und Kohle **17**, 173—178 (1964).

STOUT, J. W., ROBIE, R. A.: Heat capacity from 11 to 300° K, entropy, and heat of formation of dolomite. J. Phys. Chem. **67**, 2248—2252 (1963).

STRAKHOV, N. M. (Ed.): Types of dolomite rocks and their genesis. (in Russian). Moscow: Akad. Nauk SSSR, Trudy Geol. Inst. 1956.

STRAKHOV, N. M.: Facts and hypotheses on the question of the formation of dolomite rocks. (in Russian). Izv. Akad. Nauk SSSR, Ser. Geol. **1958**, 3—22.

SUNDIUS, N., BLIX, R.: Norsethite from Långban. Arkiv Mineral. Geol. **4**, 277—278 (1965).

SWANSON, H. E., TATGE, E.: Standard X-ray powder diffraction patterns (CaO, MgO). U.S. Natl. Bur. Standards Circ. **539**, I (1953).

SWANSON, H. E., FUYAT, R. K.: Standard X-ray diffraction powder patterns (witherite, cerussite, calcite). U.S. Natl. Bur. Standards Circ. **539**, II (1953).

SWANSON, H. E., FUYAT, R. K., UGRINIC, G. M.: Standard X-ray diffraction powder patterns (aragonite, strontianite). U.S. Natl. Bur. Standards Circ. **539**, III (1954).

SWANSON, H. E., GILFRICH, N. T., COOK, M. I.: Standard X-ray diffraction powder patterns (magnesite, rhodochrosite). U.S. Natl. Bur. Standards Circ. **539**, VII (1957).

SWANSON, H. E., GILFRICH, N. T., COOK, M. I., STINCHFIELD, R., PARKS, P. C.: Standard X-ray diffraction powder patterns (thermonatrite). U.S. Natl. Bur. Standards Circ. **539**, VIII (1959).

TAFT, W. H.: Physical chemistry of formation of carbonates. In: CHILINGAR, G. V. et al. (Eds.): Carbonate rocks, B, pp. 151—167. Amsterdam: Elsevier 1967.

TAKEDA, H., DONNAY, J. D. H.: Compound tessellations in crystal structures. Acta Cryst. **19**, 474—475 (1965).

TAN, F. C., HUDSON, J. D.: Carbon and oxygen isotopic relationships of dolomites and co-existing calcites, Great Estuarine Series (Jurassic), Scotland. Geochim. Cosmochim. Acta **35**, 755—767 (1971).

TASSEL, R. VAN: Carbonatniederschläge aus gemischten Calcium-Magnesiumchloridlösungen: Z. Anorg. Allg. Chem. **319**, 107—112 (1962).

TAYLOR, J. H.: Baryte-bearing nodules from the Middle Lias of the English east Midlands: Mineral. Mag. **29**, 18—26 (1950).

TERADA, J.: Crystal structure of the Ba, Sr and Ca triple carbonate. J. Phys. Soc. Japan **8**, 158—164 (1953).

THOMPSON, G., BOWEN, V. T., MELSON, W. G., CIFELLI, R.: Lithified carbonates from the deep-sea of the equatorial Atlantic. J. Sediment. Petrol. **38**, 1305—1312 (1968).

TRAVIS, D. F.: Structural features of mineralization from tissue to macromolecular levels of organization in the decapod crustacea. Ann. N.Y. Acad. Sci. **109**, 177—245 (1963).

UPSHAW, C. F., TODD, R. G., ALLEN, B. D.: Fluoridization of microfossils. J. Paleont. **31**, 793—795 (1957).

USDOWSKI, H. E.: Der Rogenstein des norddeutschen Unteren Buntsandsteins, ein Kalkoolith des marinen Faziesbereichs. Fortschr. Geol. Rheinland Westfalen **10**, 337—342 (1963).

USDOWSKI, H. E.: Die Genese von Dolomit in Sedimenten. Berlin-Heidelberg-New York: Springer 1967.

USDOWSKI, E.: The formation of dolomite in sediments. In: MÜLLER, FRIEDMAN (Eds.): Recent developments in carbonate sedimentology in Central Europe, pp. 21—32. Berlin-Heidelberg-New York: Springer 1968.

VAN'T HOFF, J. H.: Untersuchungen über die Bildungsverhältnisse der ozeanischen Salzablagerungen insbesondere des Stassfurter Salzlagers. Leipzig: Akad. Verlagsgesellschaft 1912.

VAN TUYL, F. M.: The origin of dolomite. Iowa Geol. Survey Ann. Rep. 1914, **25**, 251—421 (1916).

VATER, H.: Über den Einfluss der Lösungsgenossen auf die Krystallisation des Calciumcarbonats: Z. Kryst. **21**, 433—490, **22**, 209—228, **24**, 366—377, 378—404, **27**, 477—504; **30**, 295—298; 485—508; **31**, 538—578 (1893, 1894, 1895, 1897, 1899).

VATER, H.: Ueber Ktypeit und Conchit. Z. Kryst. **35**, 149—178 (1902).

VILLIERS, J. P. R. DE: Crystal structures of aragonite, strontianite, and witherite. Am. Mineralogist **56**, 758—767 (1971).

WASASTJERNA, J. A.: The crystal structure of dolomite. Soc. Sci. Fenn. Com. Phys. Math. **2**, No. 14, 1—14 (1924).

WEDEPOHL, K. H.: Die Zusammensetzung der Erdkruste. Fortschr. Mineral. **46**, 145—174 (1969).

WEDEPOHL, K. H.: Composition and abundance of common sedimentary rocks. In: Handbook of Geochemistry, Vol. I, pp. 250—271. Berlin-Heidelberg-New York: Springer 1969.

WEGSCHEIDER, R., MEHL, J.: Über Systeme Na_2CO_3-$NaHCO_3$-H_2O und das Existenzgebiet der Trona. Monatsh. Chem. **49**, 283—315 (1928).

WELLS, A. J., ILLING, L. V.: Present day precipitation of calcium carbonate in the Persian Gulf. In: VAN STRAATEN (Ed.) Deltaic and shallow marine deposits, pp. 429—435. Amsterdam: Elsevier 1964.

WEYL, P. K.: Porosity through dolomitization: conservation-of-mass requirements. J. Sediment. Petrol. **30**, 85—90 (1960).

WEYL, P. K.: The carbonate saturometer. J. Geol. **69**, 32—44 (1961).

WILBUR, K. M., WATABE, N.: Influence of the organic matrix on crystal type in molluscs. Nature **188**, 334 (1960).

WILBUR, K. M., WATABE, N.: Experimental studies on calcification in molluscs and the alga Coccolithus Huxleyi. Ann. N. Y. Acad. Sci. **109**, 82—112 (1963).

WRAY, J. L., DANIELS, F.: Precipitation of calcite and aragonite. J. Am. Chem. Soc. **79**, 2031—2034 (1957).

WYCKOFF, R. W. G.: The crystal structures of some carbonates of the calcite group. Am. J. Sci. 4th ser. **50**, 317—360 (1920).

WYCKOFF, R. W. G., MERWIN, H. E.: The crystal structure of dolomite. Am. J. Sci. 5th ser. **8**, 447—461 (1924).

ZELLER, E. J., WRAY, J. L.: Factors influencing precipitation of calcium carbonate. Bull. Am. Ass. Petrol. Geol. **40**, 140—152 (1956).

ZOBELL, C. E.: Microbial transformation of molecular hydrogen in marine sediments with particular reference to petroleum: Bull. Am. Ass. Petrol. Geol. **31**, 1709—1751 (1947).

ZOBELL, C. E.: Organic geochemistry of sulfur. In: BREGER, I. A. (Ed.): Organic Geochemistry, pp. 543—578. Oxford: Pergamon 1963.

Subject Index

Minerals, Rocks and Inorganic Materials